Universitext

Universitext

Editors (North America): J.H. Ewing, F.W. Gehring, and P.R. Halmos

Aksoy/Khamsi: Nonstandard Methods in Fixed Point Theory
Aupetit: A Primer on Spectral Theory
Bachumikern: Linear Programming Duality
Benedetti/Petronio: Lectures on Hyperbolic Geometry
Berger: Geometry I, II (two volumes)
Bliedtner/Hansen: Potential Theory
Booss/Bleecker: Topology and Analysis
Carleson/Gamelin: Complex Dynamics
Cecil: Lie Sphere Geometry: With Applications to Submanifolds
Chandrasekharan: Classical Fourier Transforms
Charlap: Bieberbach Groups and Flat Manifolds
Chern: Complex Manifolds Without Potential Theory
Cohn: A Classical Invitation to Algebraic Numbers and Class Fields
Curtis: Abstract Linear Algebra
Curtis: Matrix Groups
van Dalen: Logic and Structure
Das: The Special Theory of Relativity: A Mathematical Exposition
Devlin: Fundamentals of Contemporary Set Theory
DiBenedetto: Degenerate Parabolic Equations
Dimca: Singularities and Topology of Hypersurfaces
Edwards: A Formal Background to Mathematics I a/b
Edwards: A Formal Background to Mathematics II a/b
Emery: Stochastic Calculus
Foulds: Graph Theory Applications
Frauenthal: Mathematical Modeling in Epidemiology
Fukhs/Rokhlin: Beginner's Course in Topology
Gallot/Hulin/Lafontaine: Riemannian Geometry
Gardiner: A First Course in Group Theory
Gårding/Tambour: Algebra for Computer Science
Godbillon: Dynamical Systems on Surfaces
Goldblatt: Orthogonality and Spacetime Geometry
Hahn: Quadradic Algebras, Clifford Algebras, and Arithmetic of Forms
Hiawka/Schoissengeier/Taschner: Geometric and Analytic Number Theory
Holmgren: A First Course in Discrete Dynamical Systems
Howe/Tan: Non-Abelian Harmonic Analysis: Applications of $SL(2,R)$
Humi/Miller: Second Course in Ordinary Differential Equations
Hurwitz/Kritikos: Lectures on Number Theory
Iversen: Cohomology of Sheaves
Jennings: Modern Geometry with Applications
Jones/Morris/Pearson: Abstract Algebra and Famous Impossibilities
Kelly/Matthews: The Non-Euclidean Hyperbolic Plane
Kempf: Complex Abelian Varieties and Theta Functions
Kostrikin: Introduction to Algebra
Krasnoselskii/Pekrovskii: Systems with Hysteresis
Luecking/Rubel: Complex Analysis: A Functional Analysis Approach
MacLane/Moerdijk: Sheaves in Geometry and Logic
Marcus: Number Fields

(continued after index)

Hans Sagan

Space-Filling Curves

With 65 Figures

Springer-Verlag
New York Berlin Heidelberg London Paris
Tokyo Hong Kong Barcelona Budapest

Hans Sagan
5004 Glen Forest Drive
Raleigh, NC 27612 USA
and
Mathematisches Institut der
Universität Wien
Strudlhofgasse 4
A-1090 Wien, Austria

Mathematics Subject Classifications (1991): 02-40

Library of Congress Cataloging-in-Publication Data
Sagan, Hans. Space-filling curves / Hans Sagan.
p. cm. – (Universitext)
Includes bibliographical references and index.
ISBN 0-387-94265-3 (New York). – ISBN 3-540-94265-3 (Berlin)
1. Curves on surfaces. 2. Topology. I. Title.
QA643.S12 1994
516.3′62–dc20 94-246

Printed on acid-free paper.

Production managed by Hal Henglein; manufacturing supervised by Vincent Scelta.
Photocomposed copy prepared from the author's TeX files.
Printed and bound by R.R. Donnelley & Sons, Harrisonburg, VA.
Printed in the United States of America.

9 8 7 6 5 4 3 2 1

ISBN 0-387-94265-3 Springer-Verlag New York Berlin Heidelberg
ISBN 3-540-94265-3 Springer-Verlag Berlin Heidelberg New York

To My Grandchildren
Jesse Jo and John Byron

Preface

The subject of space-filling curves has fascinated mathematicians for over a century and has intrigued many generations of students of mathematics. Working in this area is like skating on the edge of reason. Unfortunately, no comprehensive treatment has ever been attempted other than the gallant effort by W. Sierpiński in 1912. At that time, the subject was still in its infancy and the most interesting and perplexing results were still to come. Besides, Sierpiński's paper was written in Polish and published in a journal that is not readily accessible (Sierpiński [2]). Most of the early literature on the subject is in French, German, and Polish, providing an additional *raison d'être* for a comprehensive treatment in English. While there was, understandably, some intensive research activity on this subject around the turn of the century, contributions have, nevertheless, continued up to the present and there is no end in sight, indicating that the subject is still very much alive. The recent interest in fractals has refocused interest on space-filling curves, and the study of fractals has thrown some new light on this small but venerable part of mathematics.

This monograph is neither a textbook nor an encyclopedic treatment of the subject nor a historical account, but it is a little of each.

While it may lend structure to a seminar or pro-seminar, or be useful as a supplement in a course on topology or mathematical analysis, it is primarily intended for self-study by the aficionados of classical analysis.

The long list of references that I have included may lift some of the burden of extensive literature searches off those who wish to work in this area. I cannot claim with any assurance that the list is complete, but it is more extensive than one is likely to find elsewhere.

I have tried to put the development into some historical perspective by embellishing it with direct quotations and providing short biographies and photographs of the mathematicians that have been most prominently involved in the development of the subject. For the benefit of the polyglots among my readers, I have presented the quotations in the original language to convey their authentic flavor, because I found it impossible to communicate the exact meaning and spirit in the English translations that I have provided, due to the distinct peculiarities that are characteristic of different languages. A considerable effort to find photographs of the featured mathematicians that date from the time when they made their major contribution to the subject met, regrettably, with only limited success. As the reader can see from the biographical data that have been provided and the publication dates of the pertinent articles, they were all quite young when they made their contributions. In fact, the average age was just a little above 30, old enough to have reached mathematical maturity and full command of the necessary analytical tools, and yet young enough to entertain bold and unconventional ideas.

I hope that I have included all the material that is relevant to the subject. Some of it I have treated in detail, some I have only summarized. Beyond a certain core that is obligatory in any treatment of this subject, I was guided by personal taste in the selection of topics that were to be covered in detail. It was very tempting to enlarge on some of the other aspects but, unless one draws the line somewhere, one winds up with a 400 page book that no one wants to buy or read.

To study Chapters 1 through 5, Sections 6.1 through 6.8, and Chapter 7 with understanding, the reader will need a background that is usually provided in a respectable course on Advanced Calculus. In Section 6.9 and Chapters 8 and 9, the reader is expected to be familiar with the basic notions of measure theory, and in Section 6.10, some knowledge of the theory of analytic functions (of a complex variable) is required. Chapters 2 through 5, and Chapter 7 without Section 7.5, which acquaint the reader with the prototypes of space-filling curves, may be studied quite independently of Chapters 6, 8, and 9.

In Chapter 6, I provide a direct route to the Hahn-Mazurkiewicz theorem without subjecting the reader to 200 pages of definitions and lemmas. The reader is not expected to have any background in point-set topology. I have provided all the prerequisites in Sections 6.1 through 6.7. Sacrificing elegance, I have avoided the concept of *relatively open (closed)* because it has been my experience over many years that non-topologists feel uncomfortable with this notion.

The last chapter on Fractals serves as a unifying framework for the generation of the space-filling curves of Peano, Hilbert, and Sierpiński. In addition, it provides a unique opportunity for the display of many attractive illustrations. However, when looking up from my roughly hewn writing table that comes with the rustic furnishings of my rented cabin, I behold the most beautiful fractal of them all: the skyline of the Grand Tetons in Wyoming.

Jackson, Wyoming
August 1993

H.S.

Acknowledgments

In the preparation of this manuscript, I found help, advice, valuable hints, and references from many of my colleagues. Listed alphabetically and by country, they are: Auguste Dick, Wilhelm Frank, Peter Gruber, Edmund Hlawka, and Walter Wunderlich (Austria); Gerhard Betsch, Konrad Jacobs, and B. Mund (Germany); Livia Maria Giacardi (Italy); Piotr Antosik, Jósef Krasinkiewicz, Krzysztof Loskot, Zofia Pawlikowska-Brożek, Andrzej Schinzel, and the Polish Academy of Sciences, who graciously extended their hospitality to me for a memorable week in the fall of 1992 (Poland); Federico Prat-Villar (Spain); Gerald L. Alexanderson, Robert B. Burckel, Richard E. Chandler, Gary D. Faulkner, Peter D. Lax, and Nicholas J. Rose (U.S.). The latter also provided substantial assistance and counsel in the technical preparation of this manuscript. In fact, without his selfless dedication, this manuscript, including the illustrations, may never have seen the light of day. Under his tutelage, Dava L. House typed the first six chapters of this manuscript in $\mathcal{A}_{\mathcal{M}}\mathcal{S}T_EX$, and she and Dionne R. Wilson prepared the illustrations by tracing and enhancing my hand-drawn sketches using Adobe Illustrator^{TM} on the Apple Macintosh^{R}. The $\TeX$ mavens at Springer-Verlag did the last three chapters and the Appendix to ensure publication before the end of the millenium. Thomas von Foerster, Senior Editor for Physics and Mathematics at Springer-Verlag New York, and his colleagues and staff sustained me with their encouragement and support from the inception of this project to its conclusion. I express my profound gratitude and heartfelt thanks to all of them! It goes without saying that I assume responsibility for all errors. I hope there are none but I know from experience that there are bound to be some.

Illustration Credits

The photograph of George Cantor on p. 5 is from the files of the Mathematical Intelligencer and reproduced here by permission from Springer-Verlag New York. The photo of Hans Hahn on p. 86 is from the Fotoarchiv, Mathematisches Forschungsinstitut Oberwolfach, and reprinted with their permission. The photograph of David Hilbert on p. 9 appeared in *The Pólya Picture Album* (Birkhäuser-Basel, 1987) and is reproduced here with the publisher's permission. The photo of Konrad Knopp on p. 55 comes from the Fotoarchiv, Mathematisches Forschungsinstitut Oberwolfach and is reproduced with their permission. The photo of Henri L. Lebesgue on p. 76 appeared in his *Oeuvres Scientifiques, Vol. I* and is reproduced with permission from the publishers, the Institut de Mathématiques, Université de Genève. The photo of Stefan Mazurkiewicz on p. 87 appeared in his *Travaux de Topologie et ses Applications*, published by PWN (Polish Scientific Publishers) in 1969, and is reproduced here with permission from the publishers. The photo of Eliakim H. Moore on p. 24 comes from the files of the Mathematical Intelligencer and is reproduced with the permission from Springer-Verlag New York. The photo of Eugen Netto on p. 6, which was given to the author by K. Jacobs from the University of Erlangen-Nürnberg, appeared in the special edition of *Acta Mathematica 1882–1912*, table générale des Tomes 1–35, and appears here with the permission of Arne Jensen of the Institut Mittag Leffler. The photo of William F. Osgood on p. 133 appeared in the Secretary's Report No. VII to the Class of 1886, Harvard College, 1911, and is reprinted here by permission from the Harvard University Office of News and Public Affairs. The photo of Giuseppe Peano on p. 31 was sent to the author from the Fotoarchiv Forschungsinstitut Oberwolfach. It appeared in *Bibliotheca Mathematica*, published by Umberto Allemandi & Co., and is reprinted here by permission from the publisher. The photo of George Pólya on p. 62 comes from the private collection of Andrew Lenard at Indiana University and is reprinted here with his permission. The photo of Isaac J. Schoenberg on p. 119 appeared in *The Pólya Picture Album* and is reproduced here by permission from Birkhäuser-Basel. The photo of Waclaw Sierpiński on p. 49 appeared in his *Oeuvres Choisies, Vol. 2*, published by PWN (Polish Scientific Publishers) in 1975 and is reprinted here with permission from the publishers. The photo of Walter Wunderlich on p. 43 comes from his private collection and is reproduced here with his permission.

Figure 2.1.1 on p. 10 is a facsimile of a figure that appeared in the Mathematische Annalen 38 (1891), p. 459, and is reproduced here with permission from Springer-Verlag Heidelberg. Figure 2.1.2 on p. 11 is from K. Strubecker's *Einführung in the höhere Mathematik, II*, p. 698, and is reproduced here by permission from R. Oldenbourg, München-Wien. Figure 9.5.2 on p. 162 is a reproduction from p. 502 of *Studies in Geometry* by L.M. Blumenthal and K. Menger, and is reproduced here by permission from the publishers, W.H. Freeman & Co., New York. Figure 9.6.3 on p. 165 is a facsimile of Fig. 4 in *Number Representations and Dragon Curves I* by Chandler Davis and Donald E. Knuth, J. Recreational Math. 3 (1970), p. 69, and is reproduced here by permission from the authors.

Contents

Preface vii

Acknowledgments ix

Illustration Credits xi

Chapter 1. Introduction 1

1.1. A Brief History of Space-Filling Curves 1
1.2. Notation 2
1.3. Definitions and Netto's Theorem 4
1.4. Problems 6

Chapter 2. Hilbert's Space-Filling Curve 9

2.1. Generation of Hilbert's Space-Filling Curve 9
2.2. Nowhere Differentiability of the Hilbert Curve 12
2.3. A Complex Representation of the Hilbert Curve 13
2.4. Arithmetization of the Hilbert Curve 18
2.5. An Analytic Proof of the Nowhere Differentiability of the Hilbert Curve 19
2.6. Approximating Polygons for the Hilbert Curve 21
2.7. Moore's Version of the Hilbert Curve 24
2.8. A Three-Dimensional Hilbert Curve 26
2.9. Problems 29

Chapter 3. Peano's Space-Filling Curve 31

3.1. Definition of Peano's Space-Filling Curve 31
3.2. Nowhere Differentiability of the Peano Curve 34
3.3. Geometric Generation of the Peano Curve 34
3.4. Proof that the Peano Curve and the Geometric Peano Curve are the Same 36
3.5. Cesàro's Representation of the Peano Curve 40
3.6. Approximating Polygons for the Peano Curve 42
3.7. Wunderlich's Versions of the Peano Curve 43

3.8. A Three-Dimensional Peano Curve 45
3.9. Problems 46

Chapter 4. Sierpiński's Space-Filling Curve 49
4.1. Sierpiński's Original Definition 49
4.2. Geometric Generation and Knopp's Representation of the Sierpiński Curve 51
4.3. Representation of the Sierpiński-Knopp Curve in Terms of Quaternaries 56
4.4. Nowhere Differentiability of the Sierpiński-Knopp Curve 58
4.5. Approximating Polygons for the Sierpiński-Knopp Curve 60
4.6. Pólya's Generalization of the Sierpiński-Knopp Curve 62
4.7. Problems 67

Chapter 5. Lebesgue's Space-Filling Curve 69
5.1. The Cantor Set 69
5.2. Properties of the Cantor Set 71
5.3. The Cantor Function and the Devil's Staircase 74
5.4. Lebesgue's Definition of a Space-Filling Curve 76
5.5. Approximating Polygons for the Lebesgue Curve 79
5.6. Problems 82

Chapter 6. Continuous Images of a Line Segment 85
6.1. Preliminary Remarks and a Global Characterization of Continuity 85
6.2. Compact Sets 91
6.3. Connected Sets 94
6.4. Proof of Netto's Theorem 97
6.5. Locally Connected Sets 98
6.6. A Theorem by Hausdorff 99
6.7. Pathwise Connectedness 101
6.8. The Hahn-Mazurkiewicz Theorem 106
6.9. Generation of Space-Filling Curves by Stochastically Independent Functions 108
6.10. Representation of a Space-Filling Curve by an Analytic Function 112
6.11. Problems 115

Chapter 7. Schoenberg's Space-Filling Curve 119
7.1. Definition and Basic Properties 119
7.2. The Nowhere Differentiability of the Schoenberg Curve 121

7.3. Approximating Polygons 123
7.4. A Three-Dimensional Schoenberg Curve 127
7.5. An $\aleph_0$-Dimensional Schoenberg Curve 128
7.6. Problems 129

Chapter 8. Jordan Curves of Positive Lebesgue Measure **131**
8.1. Jordan Curves 131
8.2. Osgood's Jordan Curves of Positive Measure 132
8.3. The Osgood Curves of Sierpiński and Knopp 136
8.4. Other Osgood Curves 140
8.5. Problems 142

Chapter 9. Fractals **145**
9.1. Examples 145
9.2. The Space where Fractals are Made 149
9.3. The Invariant Attractor Set 154
9.4. Similarity Dimension 156
9.5. Cantor Curves 159
9.6. The Heighway-Dragon 162
9.7. Problems 165

Appendix **169**
A.1. Computer Programs 169
A.1.1. Computation of the Nodal Points of the Hilbert Curve 169
A.1.2. Computation of the Nodal Points of the Peano Curve 170
A.1.3. Computation of the Nodal Points of the Sierpiński-Knopp Curve 171
A.1.4. Plotting Program for the Approximating Polygons of the Schoenberg Curve 172
A.2. Theorems from Analysis 173
A.2.1. Binary and Other Representations 173
A.2.2. Condition for Non-Differentiability 174
A.2.3. Completeness of the Euclidean Space 174
A.2.4. Uniform Convergence 174
A.2.5. Measure of the Intersection of a Decreasing Sequence of Sets 174
A.2.6. Cantor's Intersection Theorem 175
A.2.7. Infinite Products 175

References **177**

Index **187**

Chapter 1

Introduction

1.1. A Brief History of Space-Filling Curves

In 1878, George Cantor demonstrated that any two finite-dimensional smooth manifolds, no matter what their dimensions, have the same cardinality, and Mathematics has never been the same since. Cantor's finding implies, in particular, that the interval $[0,1]$ can be mapped bijectively onto the square $[0,1]^2$. The question arose almost immediately whether or not such a mapping can possibly be continuous. In 1879, E. Netto put an end to such speculation by showing that such a bijective mapping is necessarily discontinuous. Suppose the condition of bijectivity were dropped, is it still possible to obtain a continuous surjective mapping from $[0,1]$ onto $[0,1]^2$? Since a continuous mapping from $[0,1]$ (or any other interval, for that matter) into the plane (or space) was and, to a large extent, still is called a curve, the question may be rephrased as follows: Is there a curve that passes through every point of a two-dimensional region (such as, for example, $[0,1]^2$) with positive Jordan content (area)? G. Peano settled this question once and for all in 1890 by constructing the first such curve. Curves with this property are now called space-filling curves or Peano curves. Further examples by D. Hilbert (in 1891), E.H. Moore (in 1900), H. Lebesgue (in 1904), W. Sierpiński (in 1912), G. Pólya (in 1913), and others followed. Around the turn of the century, when many of these curves were discovered, the term *space* was primarily used for the three-dimensional space, and these curves were called surface-filling curves, as is apparent from the titles of the early papers on the subject ("Abbildung einer Linie auf ein Flächenstück" = "mapping of a line onto a piece of a surface," "Une courbe qui remplit une aire plane" = "A curve that fills a plane region," or, more specifically, "O krzywych, wypolniajacych kwadrat" = "On curves that fill a square").

However, this was not the end of it. Since it turned out to be impossible to map $[0,1]$ bijectively and continuously onto $[0,1]^2$ (or any other two-dimensional region with positive Jordan content), it was asked if it could be done onto a region with positive outer measure. In other words: Are there Jordan curves (continuous injective maps from $[0,1]$ into $\mathbb{E}^n$, $n \geq 2$), with positive n-dimensional outer measure? There are indeed, as W.F. Osgood showed in 1903 when he constructed a one-parameter family of such curves. In fact, Osgood's curves have positive Lebesgue measure, and the limiting arc of his family is Peano's space-filling curve. Other examples of such families followed, such as the one by K. Knopp in 1917 with Sierpińki's space-filling curve as limiting arc.

With the square (and all its continuous images) revealed as a continuous image of $[0, 1]$ (or any line segment, for that matter), the question arose as to the general structure of a continuous image of a line segment. In 1908, A. Schoenflies found a criterion which, by its very nature, only applies to subsets of the two-dimensional plane and never entered the mathematical mainstream. The further pursuit of this question led to the new topological concept of local connectedness and to a complete answer found independently by S. Mazurkiewicz and H. Hahn in 1913. It turned out that a set is the continuous image of a line segment if and only if it is compact, connected, and locally connected, and this criterion applies not only to the n-dimensional Euclidean space but more generally to Hausdorff spaces.

When everything seemed settled and no more surprises appeared to be lurking in the background, H. Steinhaus discovered in 1936 that if two continuous non-constant functions on $[0, 1]$ are stochastically independent with respect to Lebesgue measure, then they are the coordinate functions of a space-filling curve. He made use of this to construct space-filling curves in spaces with countably many dimensions. In 1938, I.J. Schoenberg came up with a novel modification of Lebesgue's space-filling curve, and in 1945, R. Salem and A. Zygmund constructed a lacunary power series, representing an analytic function in the open unit disk and continuous on the closed unit disk, which maps the circumference of the disk onto a square, meaning that its real and imaginary parts are the coordinate functions of a space-filling curve.

1.2. Notation

The complete ordered field of real numbers will be denoted by $\mathbb{R}$, the n-dimensional cartesian space by $\mathbb{R}^n$, and the field of complex numbers by $\mathbb{C}$.

We will use capital script letters to denote subsets of $\mathbb{R}^n$ such as

$$\mathcal{I} = [0, 1]$$

for the *closed unit-interval*,

$$\mathcal{Q} = [0, 1]^2$$

for the *closed unit-square*,

$$\mathcal{W} = [0, 1]^3$$

for the *closed unit-cube*, and $\mathcal{T}$ for the *closed triangular region* with vertices at (0,0), (2,0), (1,1).

The complement of a set $\mathcal{S}$ will be denoted by $\mathcal{S}^c$ when it is clear from the context with respect to which universal set the complement is to be taken. Otherwise, we use $\mathcal{U}\backslash\mathcal{S}$ to denote the complement of $\mathcal{S}$ with respect to $\mathcal{U}$.

Vectors will be denoted by lowercase letters and their components by lowercase greek letters such as in

$$x = \begin{pmatrix} \xi \\ \eta \end{pmatrix}, \quad \text{or} \quad x = \begin{pmatrix} \xi \\ \eta \\ \zeta \end{pmatrix}, \quad \text{or} \quad f = \begin{pmatrix} \varphi \\ \psi \end{pmatrix}, \quad \text{or} \quad f = \begin{pmatrix} \varphi \\ \psi \\ \chi \end{pmatrix}.$$

The Euclidean norm of a vector x will be denoted by

$$\|x\| = \sqrt{\xi^2 + \eta^2 (+\zeta^2)}$$

and the n-dimensional Euclidean space (which is $\mathbb{R}^n$ with the Euclidean norm defining the metric) by $\mathbb{E}^n$.

Matrices will be denoted by capital letters, H, K, P, S, with I representing the identity matrix and similarity transformations by the gothic letters $\mathfrak{F}$, $\mathfrak{H}$, $\mathfrak{K}$, $\mathfrak{L}$, $\mathfrak{P}$, and $\mathfrak{S}$.

Γ denotes the *Cantor set*, J_n the n-dimensional *Jordan content* (area, volume) of a *Jordan measurable* subset of $\mathbb{E}^n$ such as in $J_2(Q) = 1$ and Λ_n the n-dimensional *Lebesgue measure of a Lebesgue measurable* subset of $\mathbb{E}^n$ as in $\Lambda_1(\Gamma) = 0$.

An *injective* map from $\mathcal{A}$ into $\mathcal{B}$ will be denoted by

$$\mathcal{A} \xrightarrow{1-1} \mathcal{B},$$

a *surjective* map from $\mathcal{A}$ onto $\mathcal{B}$ by

$$\mathcal{A} \xrightarrow{\text{onto}} \mathcal{B},$$

and a *bijective* map from $\mathcal{A}$ onto $\mathcal{B}$ by

$$\mathcal{A} \longleftrightarrow \mathcal{B}.$$

Binaries will be denoted by

$$0_{\dot{2}} b_1 b_2 b_3 \ldots = b_1/2 + b_2/2^2 + b_3/2^3 + \cdots, \; b_j = 0 \text{ or } 1,$$

ternaries by

$$0_{\dot{3}} t_1 t_2 t_3 \ldots = t_1/3 + t_2/3^2 + t_3/3^3 + \cdots, \; t_j = 0, 1, \text{ or } 2,$$

quaternaries by

$$0_{\dot{4}} q_1 q_2 q_3 \ldots = q_1/4 + q_2/4^2 + q_3/4^3 + \cdots, \; q_j = 0, 1, 2, \text{ or } 3,$$

and *octals* by

$$0_{\dot{8}} \omega_1 \omega_2 \omega_3 \ldots = \omega_1/8 + \omega_2/8^2 + \omega_3/8^3 + \ldots, \; \omega_j = 0, 1, 2, \ldots, \text{ or } 7.$$

The *dyadic* (binary) *representation* of an integer is denoted by

$$(b_0b_1b_2\dots b_n)_2 = b_0 2^n + b_1 2^{n-1} + b_2 2^{n-2} + \dots + b_n$$

and the *triadic* representation of an integer by

$$(t_0t_1t_2\dots t_n)_3 = t_0 3^n + t_1 3^{n-1} + t_2 3^{n-2} + \dots + t_n.$$

We use upper bars to denote periods, as in

$$0_{\natural} a_1 a_2 a_3 \dots a_n \overline{b_1 b_2 b_3 \dots b_k} = 0_{\natural} a_1 a_2 a_3 \dots a_n b_1 b_2 b_3 \dots b_k b_1 b_2 b_3 \dots b_k \dots .$$

(See also Appendix A.2.1.)

$\triangleq$ means "equal by definition," and a small square $\square$ at the end of a proof stands for *quod erat demonstrandum.*

Throughout this treatment, we will be dealing with functions from a closed and bounded line segment into $\mathbb{E}^n$. Since any closed and bounded line segment is homeomorphic to $\mathcal{I}$, we may, and will, assume without loss of generality that such functions have the domain $\mathcal{I}$. The reader can easily convince himself that

$$\left.\begin{aligned} \xi &= \xi_1 + (\xi_2 - \xi_1)t \\ \eta &= \eta_1 + (\eta_2 - \eta_1)t \end{aligned}\right\} \qquad 0 \le t \le 1,\ \xi_1 \ne \xi_2 \quad \text{or} \quad \eta_1 \ne \eta_2 \tag{1.2.1}$$

represents a homeomorphism h from $\mathcal{I}$ onto the closed line segment from the point (ξ_1, η_1) to the point (ξ_2, η_2). It is, in fact, a diffeomorphism. If f happens to be differentiable, then so is the composite function $f \circ h$, and if f is not differentiable, then $f \circ h$ is not either.

1.3. Definitions and Netto's Theorem

(1.3.1) Definition. *If f is a function from a subset of $\mathbb{E}^m$ into $\mathbb{E}^n$, then*

$$f_*(\mathcal{A}) = \{f(x) \in \mathcal{R}(f) | x \in \mathcal{A} \cap \mathcal{D}(f)\},$$

where $\mathcal{A} \subseteq \mathbb{E}^m$, is called the direct image of $\mathcal{A}$ under f. $\mathcal{D}(f)$ denotes the domain, and $\mathcal{R}(f)$ the range of the function f.

The term *curve* means different things to different people. We find it convenient for our purpose to adopt the following definition:

(1.3.2) Definition. *If $f : \mathcal{I} \to \mathbb{E}^n$ is continuous, then the image $f_*(\mathcal{I})$ is called a curve. $f(0)$ is called the beginning point of the curve and $f(1)$ is called its endpoint.*

We call

$$x = f(t),\ t \in \mathcal{I}$$

a parameter representation of the curve $\mathcal{C} = f_*(\mathcal{I})$. In components

$$\left.\begin{array}{l}\xi = \varphi(t)\\ \eta = \psi(t)\end{array}\right\} \quad \textit{in} \quad \mathbb{E}^2 \qquad \textit{and} \qquad \left.\begin{array}{l}\xi = \varphi(t)\\ \eta = \psi(t)\\ \zeta = \chi(t)\end{array}\right\} \quad \textit{in} \quad \mathbb{E}^3.$$

This definition excludes curves without beginning and/or endpoints, but otherwise embraces most objects that are customarily viewed as curves, and then *some*. The "some" is what this treatise is all about. We will see in the next four chapters that there are curves that pass through every point of an n-dimensional region with positive Jordan content (area for $n = 2$, volume for $n = 3$) such as the square $\mathcal{Q}$ in $\mathbb{E}^2$ and the cube $\mathcal{W}$ in $\mathbb{E}^3$. Such curves are called *space-filling*:

(1.3.3) Definition. *If $f : \mathcal{I} \longrightarrow \mathbb{E}^n$, $n \geq 2$, is continuous and $J_n(f_*(\mathcal{I})) > 0$, then $f_*(\mathcal{I})$ is called a space-filling curve.*

As G. Cantor showed in 1878 (Cantor [1]), any two finite-dimensional smooth manifolds μ_m, μ_n, no matter what their dimensions, have the same *cardinality*, i.e., can be brought into a 1-1 correspondence. This means that there is a bijective map $f : \mu_m \longleftrightarrow \mu_n$. In particular, there are bijective maps $g : \mathcal{I} \longleftrightarrow \mathcal{Q}$, $h : \mathcal{I} \longleftrightarrow \mathcal{W}$, and $k : \mathcal{I} \longleftrightarrow \mathcal{T}$.

George Cantor (1845–1918) was born in St. Petersburg (which, for many years, was known as Leningrad) and died in Halle, Germany. He commenced his studies at the University of Zürich in 1862, and continued at the University of Berlin from 1863 to 1869, where he was very much under the influence of Karl Weierstraß. He wrote a thesis under the direction of E.E. Kummer and received his doctor's degree in 1867. He taught at the University of Halle from 1869 to 1905, where he was appointed professor in 1879. He has revolutionized mathematics and mathematical thinking with the development of set theory and his theory of the infinite. His hope of attaining a professorship at the University of Berlin was thwarted by L. Kronecker, who did not agree with his vision of transfinite numbers. (For more details, see Meschkowski [1].)

If f generates a space-filling curve, then it cannot be bijective, as E. Netto showed in 1879 (Netto [1]).

(1.3) Theorem. *(Netto): If f represents a bijective map from an m-dimensional smooth manifold μ_m onto an n-dimensional smooth manifold μ_n and $m \neq n$, then f is necessarily discontinuous.*

We will prove this theorem in Section 6.4 for the case where $\mu_m = \mathcal{I}$ and $\mu_n = \mathcal{Q}$ or $\mathcal{W}$, or $\mathcal{T}$. For the general case, we refer the reader to the literature (Greenberg [1]).

Eugen Netto (1848–1919), a contemporary of Cantor, was born in Halle and died in Gießen, Germany. Like Cantor, he studied at the University of Berlin from 1866 to 1870, where he was greatly influenced by Leopold Kronecker and Karl Weierstraß. In 1870, he received his doctor's degree under the direction of Weierstraß. He taught at the University of Strasbourg from 1879 to 1882, at the University of Berlin from 1882 to 1888, and finally settled at the University of Gießen, where he taught until his retirement in 1913. His major contributions to mathematics lie in the areas of group theory and combinatorics. (For more details see Biermann [1].)

An injective map from $\mathcal{I}$ into $\mathbb{E}^n$ is called a *Jordan curve.* More will be said about such curves in Chapter 8.

Netto's theorem implies that *there are no space-filling Jordan curves.* (To avoid possible confusion, let us advise the reader that some authors, such as W. Sierpiński and St. Mazurkiewicz, use the term *Jordan curve* for continuous maps from a line segment into $\mathbb{E}^n$. Some of *their* Jordan curves are space-filling!)

1.4. Problems

1. Show that $0_{\dot{2}}b_1b_2b_3\ldots b_nb = 0_{\dot{2}}b_1b_2b_3\ldots b_n(b-1)\overline{1}, \qquad b = 1,$

$$0_{\dot{3}}t_1t_2t_3\ldots t_nt = 0_{\dot{3}}t_1t_2t_3\ldots t_n(t-1)\overline{2}, \qquad t = 1 \text{ or } 2,$$

$$0_{\dot{4}}q_1q_2q_3\ldots q_nq = 0_{\dot{4}}q_1q_2q_3\ldots q_n(q-1)\overline{3}, \qquad q = 1,\ 2 \text{ or } 3.$$

2. Show that $0_{\dot{2}}b_1b_2b_3b_4\ldots = 0_{\dot{4}}(2b_1+b_2)(2b_3+b_4)\ldots$

$$0_{\dot{3}}t_1t_2t_3t_4\ldots = 0_{\dot{9}}(3t_1+t_2)(3t_3+t_4)\ldots$$

3. Represent $0_{\dot{2}}011001$ as a decimal and a quaternary, $0_{\dot{4}}301213$ as a binary and a decimal, and $0_{\dot{3}}2201202$ as a decimal and eneadic.

($0_{\dot{9}}e_1e_2e_3\ldots = e_1/9 + e_2/9^2 + e_3/9^3 + \cdots$, $e_j = 0, 1, 2, \ldots$, or 8, is called an *eneadic.*)

4. Express 3/8 as a binary, ternary, quaternary, octal.
5. Express $0_{\dot{2}}011$, $0_{\dot{3}}\overline{101}$, $0_{\dot{4}}\overline{3012}$, $0_{\dot{8}}6103$ as fractions.
6. Show: Every rational number has a finite or repeating binary, ternary, quaternary, octal representation and vice versa.
7. Show that $\mathcal{I}$ and $\mathcal{Q}$ have the same cardinality and that $\mathcal{I}$ and $\mathcal{W}$ have the same cardinality. (Hint: If $\mathcal{A} \longleftrightarrow \mathcal{B} \subseteq \mathcal{D}$ and $\mathcal{D} \longleftrightarrow \mathcal{C} \subseteq \mathcal{A}$, then $\mathcal{A} \longleftrightarrow \mathcal{D}$.)
8. Find functions f and g such that $f : \mathcal{I} \longleftrightarrow \mathcal{Q}$, $g : \mathcal{I} \longleftrightarrow \mathcal{W}$.
9. Find continuous mappings from the Cantor set

$$\Gamma = \{0_{\dot{3}}t_1t_2t_3\ldots | t_j = 0 \text{ or } 2,\ j = 1, 2, 3, \ldots\}$$

onto $\mathcal{Q}$ and $\mathcal{W}$.

10. Show that the mapping in (1.2.1) is a diffeomorphism from $\mathcal{I}$ onto the closed line segment from (ξ_1, η_1) to (ξ_2, η_2).
11. Let $t = t_{0\dot{3}}t_1t_2t_3\ldots \in [0, 1]$. Show that $t_j = [3^j t] - 3[3^{j-1}t]$, for $j = 0, 1, 2, 3, \ldots$, where $[\mathrm{t}]$ = largest integer $\leq t$.

Chapter 2

Hilbert's Space-Filling Curve

2.1. Generation of Hilbert's Space-Filling Curve

David Hilbert (1862–1943), who, more than anybody else, set the course for the mathematicians of the 20th century, was born in Königsberg, East Prussia (which was renamed Kaliningrad when it was incorporated into Russia in 1945) and died in Göttingen. He studied at the University of Königsberg, except for the second semester, which he spent at the University of Heidelberg, and received his doctor's degree in 1884. C.L.F. Lindemann (who, in 1882, succeeded in proving that π is transcendental) and, especially, A. Hurwitz were his most influential mentors at that time. After some postdoctoral studies in Leipzig and Paris, he returned to the University of Königsberg in 1886. In 1892, he became the successor of A. Hurwitz (who had left for Zürich), and in 1893, he succeeded to the chair that was held up to that time by Lindemann. In 1895, he followed a call to the University of Göttingen, where he taught until his retirement in 1930. He made major contributions to all areas of mathematics: Until 1893 he worked in algebraic forms, from 1894 to 1899 in algebraic number theory, from 1899 to 1903 in foundations of geometry, and from 1904 to 1909 in analysis, where he proved the existence of a solution to Dirichlet's problem, introduced the invariant integral into the field theory of the calculus of variations, and, with his treatment of integral equations, changed our concept of linear systems. His introduction of complete and separable dot-product spaces, now called Hilbert spaces, set the stage for the Riesz-Fischer theorem on the isomorphy and isometry of ℓ_2 *and* $\mathcal{L}_2(0,1)$ *and with this the ultimate answer to the question of convergence of a Fourier series. From*

1918 on, he devoted his time and energy to a study of the foundations of mathematics, where he was ultimately upstaged by Kurt Gödel, who showed, in effect, that Hilbert's objective of constructing a complete and consistent system was unattainable. (For more details, see Freudenthal [1].)

Although it was G. Peano who discovered the first space-filling curve, it was Hilbert who—to paraphrase E.H. Moore (Moore [1], p. 73)—made this phenomenon of surface-filling curves luminous to the geometric imagination. He was the first to recognize a general geometrical generating procedure that allowed the construction of an entire class of space-filling curves. The second sentence in his paper reads: "Die für eine solche Abbildung erforderlichen Functionen lassen sich in übersichtlicher Weise herstellen, wenn man sich der folgenden geometrischen Anschauung bedient." (The functions required for such a mapping can be represented in a lucid manner, if one avails oneself of the following geometric insight.) Thereupon he proceeded to promulgate the following heuristic principle: If the interval $\mathcal{I}$ can be mapped continuously onto the square $\mathcal{Q}$, then after partitioning $\mathcal{I}$ into four congruent subintervals and $\mathcal{Q}$ into four congruent subsquares, each subinterval can be mapped continuously onto one of the subsquares. Next, each subinterval is, in turn, partitioned into four congruent subintervals and each subsquare into four congruent subsquares, and the argument is repeated. If this is carried on ad infinitum, $\mathcal{I}$ and $\mathcal{Q}$ are partitioned into 2^{2n} congruent replicas for $n = 1, 2, 3 \ldots$ Hilbert has demonstrated that the subsquares can be arranged so that adjacent subintervals correspond to adjacent subsquares with an edge in common, and so that the inclusion relationships are preserved, i.e., if a square corresponds to an interval, then its subsquares correspond to the subintervals of that interval. We indicated in Fig. 2.1.1 how this process is to be carried out for the first three steps. The bold polygonal lines indicate the order in which the subsquares are to be taken in order to satisfy our requirements. (Figure 2.1.1, incidentally, is a facsimile of the original drawing in Hilbert's 1891 paper. We are reproducing it here with the permission of Springer-Verlag Heidelberg for sentimental reasons: It was the first illustration of the generation of a space-filling curve that appeared in print.) The sixth step is displayed in

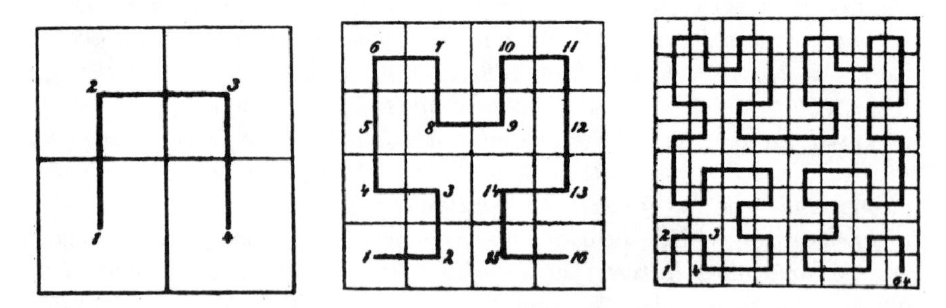

Fig. 2.1.1. Generating Hilbert's Space-Filling Curve

Fig. 2.1.2. Sixth Step in the Generation of the Hilbert Curve

Fig. 2.1.2. (We have taken this illustration from Strubecker [1], p. 698 with permission from R. Oldenbourg Verlag GmbH., München.)

We now define a mapping from $\mathcal{I}$ onto $\mathcal{Q}$ as follows:

(2.1) Definition. *Every $t \in \mathcal{I}$ is uniquely determined by a sequence of nested closed intervals (that are generated by our successive partitioning), the lengths of which shrink to 0. With this sequence corresponds a unique sequence of nested closed squares, the diagonals of which shrink into a point, and which define a unique point in $\mathcal{Q}$, the image $f_h(t)$ of t. We call $f_{h*}(\mathcal{I})$ the Hilbert Curve.*

If t is the endpoint of one of the subintervals—other than 0 or 1—then it belongs to two different sequences of nested intervals. However, since adjacent intervals are mapped onto adjacent squares, this leads to the same image.

The mapping $f_h : \mathcal{I} \longrightarrow \mathcal{Q}$ is surjective, i.e., the Hilbert curve passes through every point of the square $\mathcal{Q}$: Take any point $(\xi_0, \eta_0) \in \mathcal{Q}$. This point lies in a sequence of nested closed squares, the diagonals of which shrink into a point. With this sequence, there corresponds a sequence of nested closed subintervals the length of which shrink to 0 and, hence, define a unique $t_0 \in \mathcal{I}$, and we have $f_h(t_0) = \begin{pmatrix} \xi_0 \\ \eta_0 \end{pmatrix}$. If a point lies in a corner of a square, then it will, and if it lies on the edge of a square, then it may belong to at least two squares that do not correspond to consecutive intervals. Such a point may be viewed as belonging to two or more distinct sequences of nested closed squares and, hence, has two or more distinct preimages in $\mathcal{I}$. This should not come as a surprise in view of Netto's theorem. □

The mapping $f_h : \mathcal{I} \overset{onto}{\longrightarrow} \mathcal{Q}$ is continuous, i.e., $f_{h^*}(\mathcal{I})$ is a curve: At the nth iteration, $\mathcal{I}$ is partitioned into 2^{2n} subintervals of length $1/2^{2n}$ each. Choose $t_1, t_2 \in \mathcal{I}$ so that $|t_1 - t_2| < 1/2^{2n}$. Then, the interval $[t_1, t_2]$ overlaps with, at most, two consecutive subintervals and the images lie, at worst, in two consecutive squares of sidelength $1/2^n$ that form a rectangle with diagonal of length $\sqrt{5}/2^n$. Hence, $\| f_h(t_1) - f_h(t_2) \| \leq \sqrt{5}/2^n$ and we see that $f_h : \mathcal{I} \to \mathcal{Q}$ is continuous. □

To summarize:

(2.1) Theorem. *The Hilbert curve is a space-filling curve: $f_{h^*}(\mathcal{I}) = \mathcal{Q}$.*

2.2. Nowhere Differentiability of the Hilbert Curve

Hilbert [1] mentions *en passant* that his curve is nowhere differentiable. ("*Die oben gefundenen abbildenden Functionen sind zugleich einfache Beispiele für überall stetige and nirgends differentiirbare Functionen*" = "The mappings found above are, at the same time, simple examples of everywhere continuous and nowhere differentiable functions.") Apparently, no proof ever appeared in print until 100 years later in 1991, when this author included a simple proof in Sagan [3].

(2.2) Theorem. *The coordinate functions of the Hilbert curve $f_{h^*}(\mathcal{I})$ as defined in Definition 2.1 are nowhere differentiable.*

Proof. Let $n \geq 3$. For any $t \in \mathcal{I}$, pick a $t_n \in \mathcal{I}$ such that $|t - t_n| \leq 10/2^{2n}$ and the coordinates $\varphi_h(t), \psi_h(t)$ of the image of t are separated from the coordinates $\varphi_h(t_n), \psi_h(t_n)$ of the image of t_n by at least a square of sidelength $1/2^n$. This is always possible (see Fig. 2.1.1). Then

$$|\varphi_h(t) - \varphi_h(t_n)|/|t - t_n| \geq 2^n/10$$

An analogous argument applies to the second component ψ_h. □

2.3. A Complex Representation of the Hilbert Curve

We have been able to establish all the required properties of the Hilbert curve and its nowhere differentiability, purely on the basis of its definition by a geometric generating process. We still do not know how to calculate the coordinates of the image point of any $t \in \mathcal{I}$, other than by counting squares. However, this is only practical when t lies at the endpoint (or beginning point) of one of the 2^{2n} subintervals of the partition of $\mathcal{I}$ and n is small. Apparently, no attempt at an arithmetic-analytic representation of the Hilbert curve has been made during the past 100 years in the belief that such an attempt would be very tedious (Olmsted [1]). In 1949, Émile Borel almost succeeded when he gave an arithmetic description of the Hilbert curve ([Borel [1]). It was tedious indeed. In 1991, this author attempted and obtained a simple representation that is amenable to a numerical evaluation by computer to any desired degree of precision (Sagan [5]). It will yield the exact coordinates if $t = k/2^{2n}$, $n = 0, 1, 2, 3, \ldots$, $k = 0, 1, 2, 3, \ldots 2^{2n}$, as long as the computer's precision holds out. Preparatory to the derivation of our formula, we will establish a complex representation of the Hilbert curve (not to be confused with the representation of space-filling curves by analytic functions, which will be discussed in Section 6.10). This, in turn, is based on an idea of W. Wunderlich (Wunderlich [1], [3]), who, in the words of Karl Strubecker, has shown that Hilbert's construction may be obtained in an especially simple and perceptible manner ("*Die Hilbert'sche Konstruktion einer Peano Kurve kann nach Walter Wunderlich (1954) besonders einfach and anschaulich auf die folgende Weise erhalten werden*"–Strubecker [1], p. 697).

Looking at Fig. 2.1.1, we see that the Hilbert curve originates in the lower-left subsquare and terminates in the lower-right subsquare. Since the subsquares shrink into points, it follows that the Hilbert curve starts at the origin (entry point into $\mathcal{Q}$) and terminates at the point (1,0) (exit point from $\mathcal{Q}$). The subsquares have to be oriented in such a manner that the exit point from each subsquare coincides with the entry point of the following subsquare. In Fig. 2.3 we have indicated the correct orientation of the subsquares by bold arrows for the first two steps of the iteration process. To obtain the configuration in Fig. 2.3(b) we have to subject $\mathcal{Q}$ in Fig. 2.3(a) to the following transformations: Using complex representation, with $z \in \mathbb{C}$, we first shrink $\mathcal{Q}$ in the ratio 2:1 uniformly towards the origin: $z' = \frac{1}{2}z$. Then, we rotate the resulting square through 90° : $z'' = z'i$, and, finally, reflect on the imaginary axis: $z''' = -\overline{z}''$. Combining these three transformations and naming the composite transformation $\mathfrak{H}_0$, we obtain

$$\mathfrak{H}_0 z = \frac{1}{2}\overline{z}i. \tag{2.3.1}$$

This yields the square in the lower-left corner of Fig. 2.3(b). To obtain the square in the upper-left corner of Fig. 2.3(b), we shrink $\mathcal{Q}$ as before, and

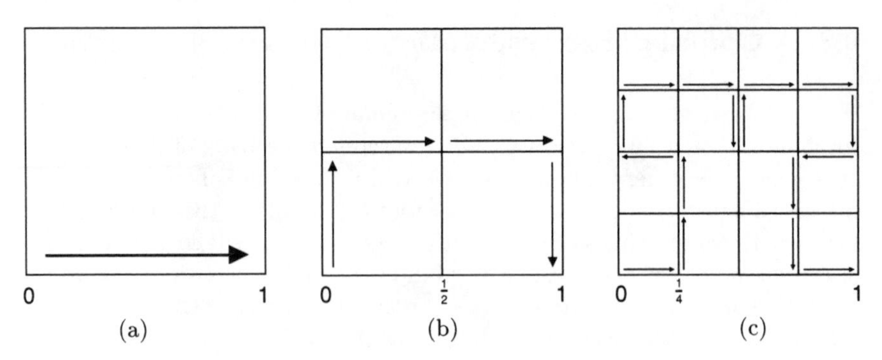

Fig. 2.3. Mapping the Square

then shift it $\frac{1}{2}$ unit upwards. This yields the transformation

$$\mathfrak{H}_1 z = \frac{1}{2}z + \frac{i}{2}. \tag{2.3.2}$$

The next square in the upper-right corner is obtained by shrinking $\mathcal{Q}$ and shifting it $\frac{1}{2}$ unit upwards and $\frac{1}{2}$ unit to the right:

$$\mathfrak{H}_2 z = \frac{1}{2}z + \frac{1}{2} + \frac{i}{2}. \tag{2.3.3}$$

Finally, we obtain the square in the lower-right corner of Fig. 2.3(b) by shrinking $\mathcal{Q}$, rotating it through $-90°(z' = -\frac{1}{2}zi)$, reflecting it on the imaginary axis $(z'' = -\overline{z}')$, and shifting the result 1 unit to the right and $\frac{1}{2}$ unit upwards to obtain

$$\mathfrak{H}_3 z = -\frac{1}{2}\overline{z}i + 1 + \frac{i}{2}. \tag{2.3.4}$$

Application of these four transformations to Fig. 2.3(a) yields Fig. 2.3(b). Application to Fig. 2.3(b) yields Fig. 2.3(c), etc. (Note that $\mathcal{Q}$ is the unique "fixed point" of the transformation $\mathfrak{H}_* : K(\mathbb{C}) \to K(\mathbb{C})$, where $K(\mathbb{C})$ is the space of all compact non-empty subsets of $\mathbb{C}$ that is defined by $\mathfrak{H}_*(\mathcal{A}) = \bigcup_{i=0}^{3} \mathfrak{H}_{i^*}(\mathcal{A})$, $\mathcal{A} \in K(\mathbb{C})$. More about this in Section 9.3.)

The nature of the generating process for the Hilbert curve suggests that it will simplify matters if we assume $t \in \mathcal{I}$ to be represented in quaternary form:

$$t = 0_4 q_1 q_2 q_3 \ldots, \quad \text{where} \quad q_j = 0, 1, 2, \text{ or } 3.$$

t lies in the $(q_1 + 1)$-th subinterval of the first partition of $\mathcal{I}$ into four subintervals. Hence, its image lies in the $(q_1 + 1)$-th square of the first partition of $\mathcal{Q}$ into four subsquares: $f(t) \in \mathfrak{H}_{q_1}\mathcal{Q}$. It lies in the $(q_2 + 1)$-th subsquare of the second partition within the $(q_1 + 1)$-th subsquare $\mathfrak{H}_{q_1}\mathcal{Q}$ of the first partition, namely $f(t) \in \mathfrak{H}_{q_1}\mathfrak{H}_{q_2}\mathcal{Q}$. We continue this reasoning ad infinitum and arrive at

$$f_h(t) = \begin{pmatrix} \mathcal{R}e \\ \mathcal{I}m \end{pmatrix} \lim_{n\to\infty} \mathfrak{H}_{q_1}\mathfrak{H}_{q_2}\mathfrak{H}_{q_3}\dots\mathfrak{H}_{q_n}\mathcal{Q}, \tag{2.3.5}$$

where the inclusion symbol $\in$ has been replaced by the equal sign, since the subsquares shrink into points.

To find the images of the finite quaternaries (endpoints or beginning points of the subintervals of the partition of $\mathcal{I}$), we first note that

$$0_{\dot{4}}q_1q_2q_3\dots q_n = 0_{\dot{4}}q_1q_2q_3\dots q_n000\dots .$$

By (2.3.5),

$$f_h(0_{\dot{4}}q_1q_2q_3\dots q_n) = \begin{pmatrix} \mathcal{R}e \\ \mathcal{I}m \end{pmatrix} \mathfrak{H}_{q_1}\mathfrak{H}_{q_2}\mathfrak{H}_{q_3}\dots\mathfrak{H}_{q_n}\mathfrak{H}_0\mathfrak{H}_0\mathfrak{H}_0\dots\mathcal{Q}.$$

Since

$$\mathfrak{H}_0\mathfrak{H}_0\mathfrak{H}_0\dots z = \lim_{n\to\infty}\mathfrak{H}_0^n z = \lim_{n\to\infty}\frac{1}{2^n}\begin{cases} i\bar{z} & \text{if n is odd} \\ z & \text{if n is even} \end{cases}$$

we have

$$\mathfrak{H}_0\mathfrak{H}_0\mathfrak{H}_0\dots z = 0$$

and, hence,

$$f_h(0_{\dot{4}}q_1q_2q_3\dots q_n) = \begin{pmatrix} \mathcal{R}e \\ \mathcal{I}m \end{pmatrix} \mathfrak{H}_{q_1}\mathfrak{H}_{q_2}\mathfrak{H}_{q_3}\dots\mathfrak{H}_{q_n}0. \tag{2.3.6}$$

Let

$$\mathfrak{H}_{q_j}z \triangleq \frac{1}{2}H_{q_j}z + \frac{1}{2}h_{q_j}, \quad j = 0, 1, 2, 3, \tag{2.3.7}$$

where $H_0z = \bar{z}i, H_1z = z, H_2z = z, H_3z = -\bar{z}i, h_0 = 0, h_1 = i, h_2 = 1+i$, and $h_3 = 2+i$. By successive substitution of H_j, h_j into (2.3.6), we obtain, in view of $H_{q_1}H_{q_2}\dots H_{q_n}0 = 0$ and with $H_{q_0}z = z$, that

$$\begin{aligned}
\mathfrak{H}_{q_1}\mathfrak{H}_{q_2}\mathfrak{H}_{q_3}\dots\mathfrak{H}_{q_n}0 &= \frac{1}{2}H_{q_1}\{\mathfrak{H}_{q_2}\mathfrak{H}_{q_3}\dots\mathfrak{H}_{q_n}0\} + \frac{1}{2}h_{q_1} \\
&= \frac{1}{2}H_{q_1}\left\{\frac{1}{2}H_{q_2}\left[\mathfrak{H}_{q_3}\dots\mathfrak{H}_{q_n}0\right] + \frac{1}{2}h_{q_2}\right\} + \frac{1}{2}h_{q_1} \\
&= \frac{1}{2}H_{q_1}\left\{\frac{1}{2}H_{q_2}\left[\frac{1}{2}H_{q_3}(\mathfrak{H}_{q_4}\dots\mathfrak{H}_{q_n}0) + \frac{1}{2}h_{q_3}\right] + \frac{1}{2}h_{q_2}\right\} + \frac{1}{2}h_{q_1} \\
&= \dots \\
&= \left(\frac{1}{2^n}\right)H_{q_1}H_{q_2}H_{q_3}\dots H_{q_n}0 + \left(\frac{1}{2^n}\right)H_{q_1}H_{q_2}H_{q_3}\dots H_{q_{n-1}}h_{q_n} \\
&\quad + (1/2^{n-1})H_{q_1}H_{q_2}H_{q_3}\dots H_{q_{n-2}}h_{q_{n-1}} + \dots + (1/2)h_{q_1} \\
&= \sum_{j=1}^{n}\left(\frac{1}{2^j}\right)H_{q_0}H_{q_1}H_{q_2}H_{q_3}\dots H_{q_{j-1}}h_{q_j}.
\end{aligned}$$

Hence,

$$f_h(0_4\dot{}q_1q_2q_3\ldots q_n) = \begin{pmatrix}\mathcal{R}e\\ \mathcal{I}m\end{pmatrix}\sum_{j=1}^{n}\left(\frac{1}{2^j}\right)H_{q_0}H_{q_1}H_{q_2}\ldots H_{q_{j-1}}h_{q_j}$$

and, in view of the continuity of f_h,

$$f_h(0_4\dot{}q_1q_2q_3\ldots) = \begin{pmatrix}\mathcal{R}e\\ \mathcal{I}_m\end{pmatrix}\sum_{j=1}^{\infty}\left(\frac{1}{2^j}\right)H_{q_0}H_{q_1}H_{q_2}\ldots H_{q_{j-1}}h_{q_j}. \tag{2.3.8}$$

In view of $H_1z = H_2z = z, H_0^2z = H_3^2z = z$, we may omit all factors

$$H_1, H_2, H_0^2 \quad \text{and} \quad H_3^2.$$

If we also note that $H_0H_3z = H_3H_0z$, we wind up with the simple formula

$$f_h(0_4\dot{}q_1q_2q_3\ldots) = \begin{pmatrix}\mathcal{R}e\\ \mathcal{I}m\end{pmatrix}\sum_{j=1}^{\infty}\left(\frac{1}{2^j}\right)H_0^{e_{0j}}H_3^{e_{3j}}h_{q_j}, \tag{2.3.9}$$

where

$$e_{kj} = \text{number of k's preceding} \quad q_j \pmod 2 \tag{2.3.9a}$$

for $k = 0$ or 3.

Although this is an expansion in terms of powers of $\frac{1}{2}$, it is not the binary representation of the image point because some coefficients may be negative.

Since a finite quaternary may also be written as an infinite quaternary with repeating 3's,

$$0_4\dot{}q_1q_2q_3\ldots q_nq = 0_4\dot{}q_1q_2q_3\ldots q_n(q-1)\overline{3},$$

we have to make sure that (2.3.9) yields the same result either way. We have from (2.3.9)

$$f_h(0_4\dot{}q_1q_2q_3\ldots q_nq) = \begin{pmatrix}\mathcal{R}e\\ \mathcal{I}m\end{pmatrix}\left\{\sum_{j=1}^{n}\left(\frac{1}{2^j}\right)H_0^{e_{0j}}H_3^{e_{3j}}h_{q_j}\right.$$
$$\left. + \left(\frac{1}{2^{n+1}}\right)H_0^{e_{0,n+1}}H_3^{e_{3,n+1}}h_q\right\}$$

and

$$f_h(0_4\dot{}q_1q_2q_3\ldots q_n(q-1)\overline{3}) = \begin{pmatrix}\mathcal{R}e\\ \mathcal{I}m\end{pmatrix}\left\{\sum_{j=1}^{n}\left(\frac{1}{2^j}\right)H_0^{e_{0j}}H_3^{e_{3j}}h_{q_j}\right.$$
$$+ \left(\frac{1}{2^{n+1}}\right)H_0^{e_{0,n+1}}H_3^{e_{3,n+1}}h_{q-1}$$
$$\left. + \left(\frac{1}{2^{n+2}}\right)H_0^{e_{0,n+1}}H_3^{e_{3,n+1}}H_{q-1}\left(h_3 + \frac{1}{2}H_3h_3 + \frac{1}{4}h_3 + \frac{1}{8}H_3h_3 + \cdots\right)\right\}.$$

Since

$$\left(I + \frac{1}{2}H_3 + \frac{1}{4}I + \frac{1}{8}H_3 + \cdots\right) h_3 = 2$$

and

$$h_q = h_{q-1} + \frac{1}{2}H_{q-1}(2)$$

for $q = 1, 2, 3$, we have

$$f_h(0_4 q_1 q_2 q_3 \ldots q_n q) = f_h(0_4 q_1 q_2 q_3 \ldots q_n (q-1)\overline{3}).$$

From (2.3.1) to (2.3.4),

$$\begin{aligned} H_0 z &= i\overline{z} = \exp(i\pi/2)\overline{z} \\ H_1 z &= z \\ H_2 z &= z \\ H_3 z &= -i\overline{z} = \exp(-i\pi/2)\overline{z} \end{aligned} \tag{2.3.10}$$

and

$$\begin{aligned} h_n &= \operatorname{sgn}(n)[(n-1) + i] \\ &= \operatorname{sgn}(n)\sqrt{1+(n-1)^2}\exp(i \ \arcsin[1/\sqrt{1+(n-1)^2}]}, \end{aligned} \tag{2.3.11}$$

where $\operatorname{sgn}(n) = \left\{\begin{matrix} 0 & \text{if} & n = 0 \\ 1 & \text{if} & n > 0 \end{matrix}\right\}$. From (2.3.10)

$$H_3^{e_{3j}} h_{q_j} = \operatorname{sgn}(q_j)\exp(-i\pi e_{3j}/2)[(q_j - 1)^2 + (-1)^{e_{3j}} i]$$

and, with

$$d_j = e_{0j} + e_{3j} \pmod 2, \tag{2.3.12}$$

where e_{kj} is defined in (2.3.9a),

$$\begin{aligned} H_0^{e_{0j}}(H_3^{e_{3j}} h_{q_j}) &= (-1)^{e_{3j}}\operatorname{sgn}(q_j)\exp[i\pi d_j/2][(q_j - 1) + (-1)^{d_j} i] \\ &= (-1)^{e_{3j}}\operatorname{sgn}(q_j)\sqrt{(q_j-1)^2+1]} \\ &\quad \times \exp[i\pi d_j/2 + i(-1)^{d_j}\arcsin(1/\sqrt{(q_j-1)^2+1)}]. \end{aligned}$$

Substitution into (2.3.9) yields the following complex representation (parametrization) of the Hilbert curve:

$$\begin{aligned} &\varphi_h(0_4 q_1 q_2 q_3 \ldots) + i\psi_h(0_4 q_1 q_2 q_3 \ldots) \\ &\quad = \sum_{j=1}^{\infty} \frac{(-1)^{e_{3j}}}{2^j}\operatorname{sgn}(q_j)\sqrt{(q_j-1)^2+1)} \\ &\quad \times \exp[i\pi d_j/2 + i(-1)^{d_j}\arcsin(1/\sqrt{(q_j-1)^2+1)}], \end{aligned} \tag{2.3.13}$$

where d_j is defined in (2.3.12).

2.4. Arithmetization of the Hilbert Curve

Decomposition of the transformations (2.3.1) to (2.3.4) into real and imaginary parts yields the following catalogue of similarity transformations:

$$\begin{aligned}
\mathfrak{H}_0\begin{pmatrix}\xi\\ \eta\end{pmatrix} &= \frac{1}{2}\begin{pmatrix}0 & 1\\ 1 & 0\end{pmatrix}\begin{pmatrix}\xi\\ \eta\end{pmatrix} + \frac{1}{2}\begin{pmatrix}0\\ 0\end{pmatrix} \triangleq \frac{1}{2}H_0\begin{pmatrix}\xi\\ \eta\end{pmatrix} + \frac{1}{2}h_0\\
\mathfrak{H}_1\begin{pmatrix}\xi\\ \eta\end{pmatrix} &= \frac{1}{2}\begin{pmatrix}1 & 0\\ 0 & 1\end{pmatrix}\begin{pmatrix}\xi\\ \eta\end{pmatrix} + \frac{1}{2}\begin{pmatrix}0\\ 1\end{pmatrix} \triangleq \frac{1}{2}H_1\begin{pmatrix}\xi\\ \eta\end{pmatrix} + \frac{1}{2}h_1\\
\mathfrak{H}_2\begin{pmatrix}\xi\\ \eta\end{pmatrix} &= \frac{1}{2}\begin{pmatrix}1 & 0\\ 0 & 1\end{pmatrix}\begin{pmatrix}\xi\\ \eta\end{pmatrix} + \frac{1}{2}\begin{pmatrix}1\\ 1\end{pmatrix} \triangleq \frac{1}{2}H_2\begin{pmatrix}\xi\\ \eta\end{pmatrix} + \frac{1}{2}h_2\\
\mathfrak{H}_3\begin{pmatrix}\xi\\ \eta\end{pmatrix} &= \frac{1}{2}\begin{pmatrix}0 & -1\\ -1 & 0\end{pmatrix}\begin{pmatrix}\xi\\ \eta\end{pmatrix} + \frac{1}{2}\begin{pmatrix}2\\ 1\end{pmatrix} \triangleq \frac{1}{2}H_3\begin{pmatrix}\xi\\ \eta\end{pmatrix} + \frac{1}{2}h_3.
\end{aligned} \tag{2.4.1}$$

These are the transformation formulas that Wunderlich has presented (Wunderlich [1] and [3]; see also Strubecker [1], p. 697 ff.). We have used the same symbol for the transformations in the complex plane as for the corresponding ones in $\mathbb{E}^2$, since they accomplish the same objective, even though they are defined by different formulas. We observe that, in the new interpretation, we still have

$$\lim_{n\to\infty}\mathfrak{H}_0^n\mathcal{Q} = \lim_{n\to\infty}\frac{1}{2^n}H_0^n\mathcal{Q} = \lim_{n\to\infty}\frac{1}{2^n}\begin{pmatrix}0 & 1\\ 1 & 0\end{pmatrix}^n\mathcal{Q} = \begin{pmatrix}0\\ 0\end{pmatrix}$$

and (2.3.9) remains unchanged:

$$f_h(0_4q_1q_2q_3\ldots) = \sum_{j=1}^{\infty}\left(\frac{1}{2^j}\right)H_0^{e_{0j}}H_3^{e_{3j}}h_{q_j}. \tag{2.4.2}$$

Checking out all the possibilities $(e_{0j}, e_{3j}) = (0,0), (1,0), (0,1)$, and $(1,1)$ for each $q_j = 1, 2$, and 3, and observing that $q_j = 0$ leads to the term $H_0^{e_{0j}}H_3^{e_{3j}}h_0 = \begin{pmatrix}0\\ 0\end{pmatrix}$, one easily verifies that (2.4.2) may be written in terms of the quaternary digits q_j as follows:

$$f_h(0_4q_1q_2q_3\ldots) = \sum_{j=1}^{\infty}(1/2^j)(-1)^{e_{0j}}\operatorname{sgn}(q_j)\begin{pmatrix}(1-d_j)q_j - 1\\ 1 - d_jq_j\end{pmatrix}, \tag{2.4.3}$$

where e_{kj} is defined in (2.3.9a) and d_j is defined in (2.3.12). This formula may also be derived directly from (2.3.13). Note that $e_{3j} + d_j = e_{0j}$ (mod 2). Formula (2.4.3) is obtained from (4) in Sagan [5] by the formal change $d_j = 1 - \delta_j$, where $\delta_j = 1$ if $e_{0j} = e_{3j}, \delta_j = 0$ if $e_{0j} \neq e_{3j}$.

Since the computation of the numerators of the terms in the sum of (2.4.3) only involves the multiplication and subtraction of integers, we can program a computer to yield the exact images of all finite $t = 0_4q_1q_2q_3\cdots q_n$

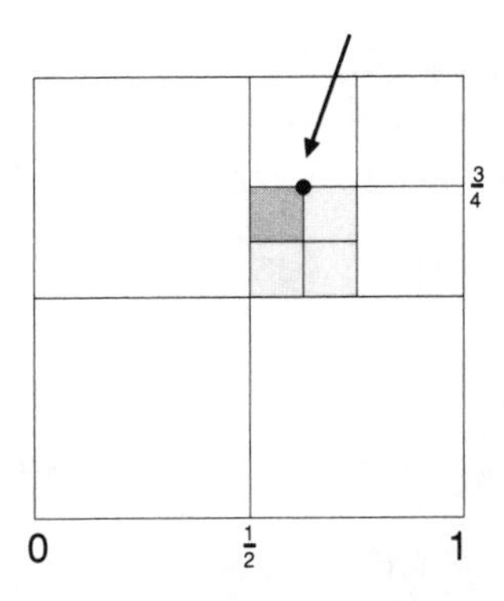

Fig. 2.4. Image of $0_4 203$

in the form of fractions with 2^n in the denominator—as long as the computer's precision holds out. By the same token, we can, in principle, calculate approximations to the coordinates of the image for any $t \in \mathcal{I}$ to any degree of precision. In Appendix A.1.1 (Program 1) we list a computer program for the evaluation of (2.4.3) in BASIC, which the reader may wish to transcribe into his or her favorite programming language.

To illustrate the use of (2.4.3), let us find $f_h(0_4 203)$. From (2.3.9a), $e_{01} = 0, e_{02} = 0, e_{03} = 1, e_{31} = 0, e_{32} = 0, e_{33} = 0$ and from (2.3.12), $d_1 = 0, d_2 = 0, d_3 = 1$. Hence,

$$\begin{aligned} f_h(0_4 203) &= (1/2)(-1)^0 \begin{pmatrix} 1.2 - 1 \\ 1 \end{pmatrix} + 0 + (1/8)(-1)^1 \begin{pmatrix} 0 - 1 \\ 1 - 1.3 \end{pmatrix} \\ &= \begin{pmatrix} 1/2 \\ 1/2 \end{pmatrix} + \begin{pmatrix} 1/8 \\ 1/4 \end{pmatrix} = \begin{pmatrix} 5/8 \\ 3/4 \end{pmatrix}. \end{aligned}$$

This represents the entry point into the fourth subsquare of the third partition within the first subsquare of the second partition within the third subsquare of the first partition. (See Fig. 2.4).

2.5. An Analytic Proof of the Nowhere Differentiability of the Hilbert Curve

In Section 2.2 we proved the nowhere differentiability of the Hilbert curve. Our proof relied, to some extent, on geometric insight. For those who might find this objectionable, we will now present an analytic proof (Sagan [9]) based on the representation of the Hilbert curve in (2.4.3):

(2.5) Theorem. *The coordinate functions of Hilbert's space-filling curve as represented in (2.4.3) are nowhere differentiable.*

Proof. We will carry out the proof for the first component φ_h. The proof for ψ_h runs along the same lines. We introduce the following abbreviating notation for the first component of the vector under the sum of (2.4.3):

$$d(d_j, q_j) = (1 - d_j)q_j - 1.$$

Then,

$$d(1, q_j) = -1 \quad \text{for} \quad q_j = 1, 2, 3; \; d(0,1) = 0, \; d(0,2) = 1, \; d(0,3) = 2. \tag{2.5.1}$$

Let $t = 0_4 q_1 q_2 q_3 \ldots \in [0,1]$. We have to distinguish three cases:

(i) There are infinitely many 1's or 2's in the quaternary representation of t. Let q_{k_1} represent the first 1 or 2 after the quaternary point, q_{k_2} the second 1 or 2, etc. ... and let

$$\underline{t}_n = 0_4 q_1 q_2 q_3 \ldots q_{k_n - 1} 0, \; \overline{t}_n = 0_4 q_1 q_2 q_3 \ldots q_{k_n - 1} 3.$$

Then

$$\overline{t}_n - \underline{t}_n = 3/4^{k_n}.$$

By (2.4.3) and (2.5.1),

$$|\varphi_h(\overline{t}_n) - \varphi_h(\underline{t}_n)| = |(-1)^{e_{0k_n}} d(d_{k_n}, 3)/2^{k_n}| \geq 1/2^{k_n}.$$

Hence,

$$|(\varphi_h(\overline{t}_n) - \varphi_h(\underline{t}_n))/(\overline{t}_n - \underline{t}_n)| = 2^{k_n}/3 \longrightarrow \infty \quad \text{as} \quad n \longrightarrow \infty,$$

i.e., φ_h is not differentiable at t.

(ii) The quaternary representation of t contains infinitely many 3's. This includes all finite quaternary representations (other than 0) which are to be replaced by infinite representations with infinitely many trailing 3's. In particular, $1 = 0_4\overline{3}$.

Let q_{k_1} represent the first digit 3 that appears after the quaternary point and is preceded by an even number of 0's or 3's, let q_{k_2} represent the second digit 3 that appears after the quaternary point and is preceded by an even number of 0's or 3's, etc.... With

$$t = 0_4 q_1 q_2 q_3 \ldots q_{k_n - 1} 3 q_{k_n + 1} \cdots,$$

let

$$t_{k_n} = 0_4 q_1 q_2 q_3 \ldots q_{k_n - 1} 0.$$

Then,

$$t - t_{k_n} = 0_4 000 \ldots 03 q_{k_n + 1} \cdots \leq 1/4^{k_n - 1}$$

and, by (2.4.3),

$$\varphi_h(t) - \varphi_h(t_{k_n}) = (-1)^{e_{0k_n}}/2^{k_n-1}$$
$$+ \sum_{m=1}^{\infty} [(-1)^{e_{0k_n+m}}/2^{k_n+m}]\operatorname{sgn}(q_{k_n+m})d(d_{k_n+m}, q_{k_n+m})$$
$$= (-1)^{e_{0k_n}}/2^{k_n-1} + e_{0m}(-1)^{e_{0k_n}}/2^{k_n}] \sum_{m=1}^{\infty} [(-1)^{e_{0m}}/2^m]$$
$$\times \operatorname{sgn}(q_{k_n+m})d(d_m, q_{k_n+m})$$
$$= (-1)^{e_{0k_n}}/2^{k_n-1} + [(-1)^{e_{0k_n}}/2^{k_n}]\varphi_h(0_4 q_{k_n+1} q_{k_n+2} q_{k_n+3} \ldots).$$

Since $|\varphi_h(t)| \leq 1$, we obtain

$$|\varphi_h(t) - \varphi_h(t_{k_n})| \geq 1/2^{k_n-1} - 1/2^{k_n} = 1/2^{k_n}.$$

Hence,

$$|(\varphi_h(t) - \varphi_h(t_{k_n}))/(t - t_{t_{k_n}})| \geq 2^{k_n-2} \longrightarrow \infty \quad \text{as} \quad n \longrightarrow \infty$$

and φ_h is not differentiable at t.

(iii) $t = 0$. Let $t_n = 3/4^n$. Then, $|0 - t| = 3/4^n$ and

$$|\varphi_h(0) - \varphi_h(t_n)| = |[(-1)^{e_{0n}}/2^n]d(d_n, 3)| \geq 1/2^n.$$

Hence,

$$|(\varphi_h(0) - \varphi_h(t_n))/(0 - t_n)| \geq 2^n/3 \longrightarrow \infty \quad \text{as} \quad n \longrightarrow \infty$$

and φ_h is not differentiable at 0.

This exhausts all possibilities, and the theorem is proved.

2.6. Approximating Polygons for the Hilbert Curve

Within each subsquare in Fig. 2.1.1, the distance between the points on the Hilbert curve and the points on the part of the bold polygonal line that lies in the subsquare is bounded above by the length of the diagonal of the subsquare. If one extends these polygons continuously by line segments from their beginning point to the origin, and from their endpoint to the point $(0, 1)$, then they approximate the Hilbert curve uniformly on $\mathcal{I}$ and it is justified to call them approximating polygons as some authors have done (such as W. Wunderlich [3]). E.H. Moore [1], by contrast, calls the polygonal lines that join the *nodes (or nodal points)* i.e., the images of the partition points of $\mathcal{I}$, approximating polygons and we have adopted the same terminology in Sagan ([5] and [6]). Specifically:

(2.6) Definition. *The polygonal line that joins the points*

$$f_h(0), f_h(1/2^{2n}), f_h(2/2^{2n}), f_h(3/2^{2n}), \ldots, f_h((2^{2n}-1)/2^{2n}), f_h(1),$$

is called the nth approximating polygon for the Hilbert curve.

In Fig. 2.6.1 we have drawn the first three approximating polygons. Since these polygons, by their nature, bump into themselves and double back now and then, it is difficult to follow their progression. To remedy the situation, we have placed arrows next to the individual segments of the approximating polygons. (The arrows are located inside the subsquares within which the Hilbert curve is approximated by the adjacent line segment.)

We may parametrize the approximating polygons $p_n : \mathcal{I} \to \mathcal{Q}$ as follows:

$$p_n(t) = 2^{2n}(t - \frac{k}{2^{2n}})f_h\left(\frac{k+1}{2^{2n}}\right) - 2^{2n}\left(t - \frac{k+1}{2^{2n}}\right)f_h\left(\frac{k}{2^{2n}}\right), \quad (2.6.1)$$

for $k/2^{2n} \leq t \leq (k+1)/2^{2n}, k = 0, 1, 2, 3, \ldots 2^{2n} - 1$, and we can say that

$$\|f_h(t) - p_n(t)\| \leq \sqrt{2}/2^n \quad \text{for all} \quad t \in \mathcal{I},$$

i.e., $\{p_n\}$ converges uniformly to the Hilbert curve. Since the p_n are continuous, this yields an alternate proof for the continuity of the Hilbert curve. (See also Appendix A.2.4.) The same may be accomplished with the approximating polygons we mentioned in the beginning of this section, which are formed from the polygonal lines in Fig. 2.1.1.

Since the square $\mathcal{Q}$ is partitioned into 2^{2n} congruent subsquares, each of sidelength $1/2^n$, and since each subsquare contains one segment of the nth approximating polygon, we obtain 2^n for the length of the nth approximating polygon. If it were meaningful to assign a length ℓ to the Hilbert

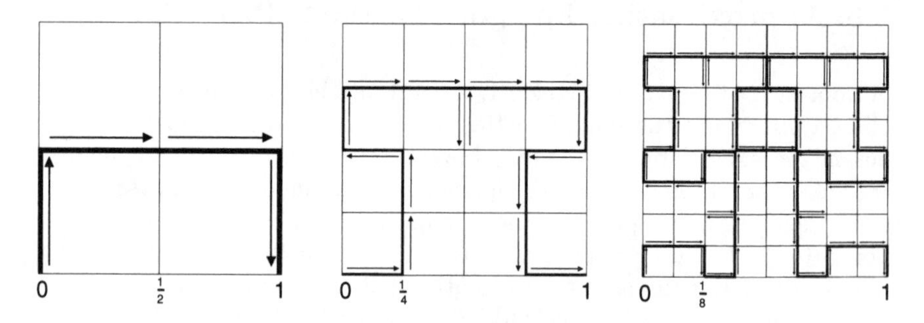

Fig. 2.6.1. Approximating Polygons for the Hilbert Curve

curve, then $\ell > 2^n$ for all n. Still, interpreting the parameter t as time, it is possible to run through the entire Hilbert curve in one time unit. This is not a contradiction, because the Hilbert curve is nowhere differentiable and, hence, it is meaningless to assign a velocity to this "motion."

The approximating polygons as defined in (2.6.1) and depicted in Fig. 2.6.1 for $n = 1, 2, 3$, may be obtained recursively from the (directed) line segment from (0,0) to (1,0) by repeated application of the transformations (2.4.1). In the limit, an $n \to \infty$, the Hilbert curve is obtained. Point sets that are obtained recursively in such a manner are loosely referred to as *fractals*. So, the Hilbert curve may be viewed as a fractal with the line segment from (0,0) to (1,0) as an *initial set*. We find *Leitmotiv* a more fitting and descriptive name for the initial set. (See also Chapter 9.)

Observe that any curve that enters the square $\mathcal{Q}$ at (0,0) and exits at (1,0) may serve as a Leitmotiv for the construction of approximating curves. In Fig. 2.6.2 we have drawn the second approximating curve with an isosceles triangle as Leitmotiv, and in Fig. 2.6.3 the third approximating curve with a semicircle as Leitmotiv.

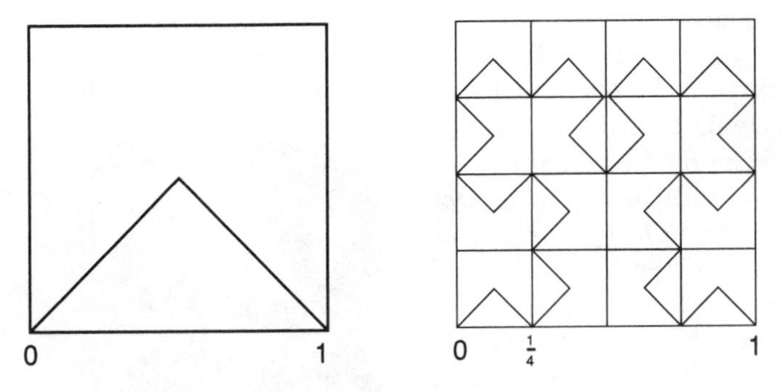

Fig. 2.6.2. Leitmotiv and Second Approximating Curve

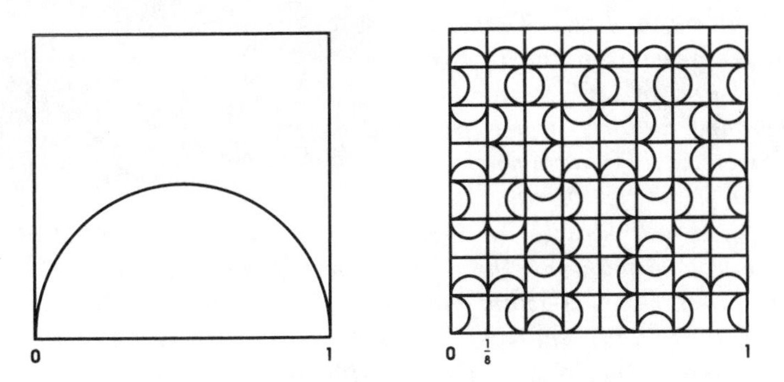

Fig. 2.6.3. Leitmotiv and Third Approximating Curve

2.7. Moore's Version of the Hilbert Curve

We are able to define a space-filling curve from $\mathcal{I}$ onto $\mathcal{Q}$ in terms of successive partitions of $\mathcal{I}$ and $\mathcal{Q}$ because it is possible to partition $\mathcal{Q}$ into 2^{2n} replicas of it, and line up the subsquares so that adjacent subintervals are mapped onto adjacent subsquares, with an edge in common so that the nth mapping preserves the $(n-1)$-th mapping. It is then possible to define each function value by utilizing the nested interval property of points in $\mathbb{E}$ and the nested square property of points in $\mathbb{E}^2$. This suggests that other space-filling curves may be obtained from the same general principle, as long as it is possible to dissect a plane figure into a number of replicas (see Gardner [2] and Golomb [1]) which can then be aligned in the required manner. We will deal with the case where the square is dissected into nine subsquares in Sections 3.1 to 3.7, and where a right isosceles triangle is dissected into two replicas in Sections 4.1 to 4.5. In Section 4.6 we will finally deal with a case where a right, but not isosceles, triangle is dissected into two similar, but not congruent, triangles. In Problem 2.2, we ask the reader to construct a space-filling curve by dissecting a certain trapezoid into four congruent replicas.

Moore [1] has shown that Hilbert's is not the only way of lining up the subsquares after successive partitionings into 2^{2n} replicas. (Except for rotations and reflections, Hilbert's is the only way if one wants the endpoint of the curve to be one unit horizontally or one unit vertically away from the beginning point.)

Eliakim Hastings Moore (1862–1932) was born in Marietta, Ohio and died in Chicago. He studied at Yale under Hubert A. Newton and received his doctor's degree in 1885. Newton, deeply impressed with Moore's abilities, financed him for a year at the universities of Göttingen and Berlin. In Göttingen, he met H. Weber, H.A. Schwarz and F. Klein; in Berlin, K. Weierstraß and L. Kronecker. (Later, in 1899, he received an honorary doctor's degree from the University of Göttingen.) He joined the faculty at Northwestern University as an instructor in 1886. He then went back to Yale University, where he worked for a year as a tutor. In 1889 he returned to Northwestern University as an assistant professor. He was promoted to an associate professorship in 1891 and, in 1892, went to the University of Chicago as professor and acting head of the department of mathematics. He became the permanent head in 1896, and retained this position until his retirement

in 1931. His major contributions to mathematics are in geometry, algebra, group theory, number theory, and analysis (integral equations). Not only was he instrumental in shaping the teaching of mathematics at the University of Chicago but, well beyond that, had great influence in the development of American mathematics. He brought Oskar Bolza to Chicago, who founded the American School of the Calculus of Variations, which brought forth such prominent scholars as G. Bliss, M. Hestenes and E.J. McShane. Among the long list of Moore's Ph.D. students, one finds such prominent names as L.E. Dickson, D.N. Lehmer, O. Veblen, and G.D. Birkhoff. In 1894, he was influential in the transformation of the local New York Mathematical Society into the American Mathematical Society. From 1899 to 1907 he was one of the chief editors of the Transactions of the American Mathematical Society, the first volume of which contains the paper we referred to above, and to which we will return in Sections 3.2 and 3.3. (For more details, see Calinger [1] and Bliss and Dickson [1].)

Moore has indicated an enumeration of the subsquares depicted for the first four steps in Fig. 2.7. Note that the space-filling curve obtained

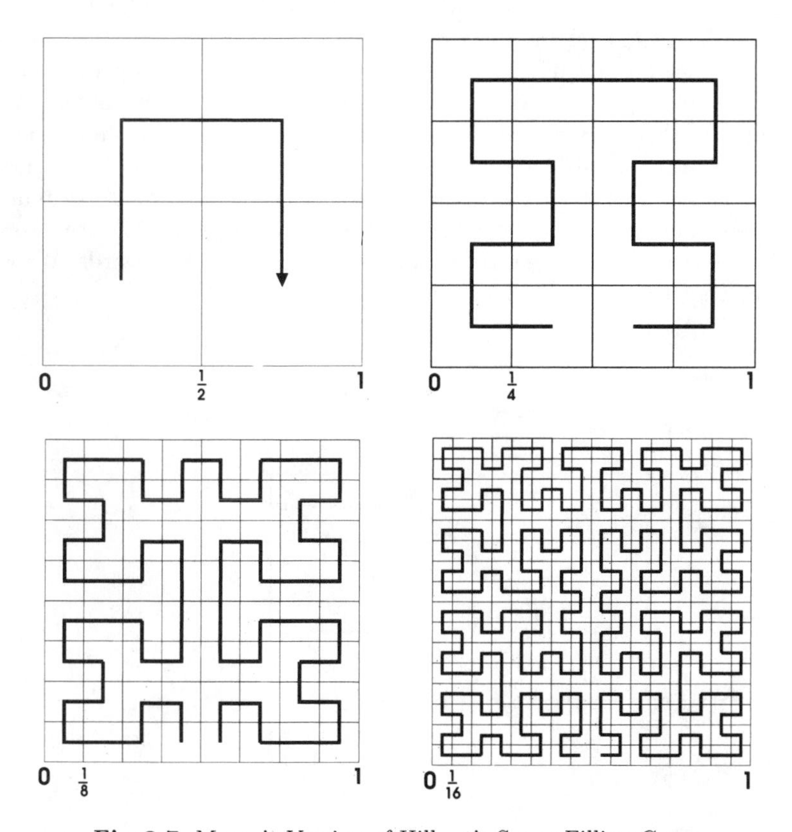

Fig. 2.7. Moore's Version of Hilbert's Space-Filling Curve

in this manner emanates from $\left(\frac{1}{2},0\right)$ and terminates at the same point, in contrast to the Hilbert curve that emanates from the origin and terminates at (1,0).

2.8. A Three-Dimensional Hilbert Curve

To define a three-dimensional Hilbert curve we proceed as in Section 2.1, except that we now divide $\mathcal{I}$ at each step into 2^{3n} congruent subintervals and, accordingly, the unit cube $\mathcal{W}$ into 2^{3n} congruent subcubes. (See also Sagan [8].) At the first step, we line up the subcubes, as indicated in Fig. 2.8.1. The polygonal line joins the midpoints of the eight subcubes in the desired order. The next step is illustrated in Fig. 2.8.2, where we have drawn the images of the clamp-line polygon from Fig. 2.8.1 with bold lines and the connectors with thin lines. (We have omitted the contours of the square and the subsquares and the coordinate axes to avoid overloading the figure.)

This process satisfies our requirements for the definition of a space-filling curve: adjacent subintervals mapped onto adjacent subcubes with a common face and each mapping preserving the preceding one. We can now define a space-filling curve by means of the nested interval property and the nested cube property, as we did in Section 2.1. The endpoint of the curve obtained in this manner lies one unit vertically above the beginning point.

In order to obtain an arithmetic-analytic representation of this curve, we proceed as in Sections 2.3 and 2.4. We place the coordinate system so that the origin coincides with the left lower-front corner of $\mathcal{W}$, the ξ-axis points to the right, the η-axis backwards, and the ζ-axis upwards. We will assume that the point $t \in \mathcal{I}$ is represented as an octal

$$t = 0_{\dot{8}}\omega_1\omega_2\omega_3\ldots, \quad \text{where} \quad \omega_j = 0,1,2,3,4,5,6, \text{ or } 7.$$

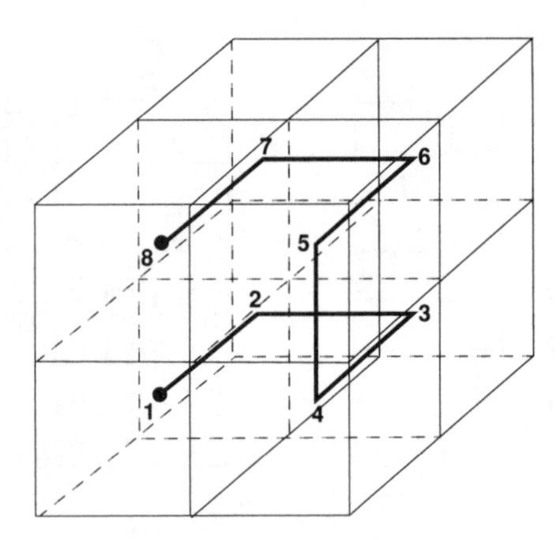

Fig. 2.8.1. First Step in the Generation of a Three-Dimensional Hilbert Curve

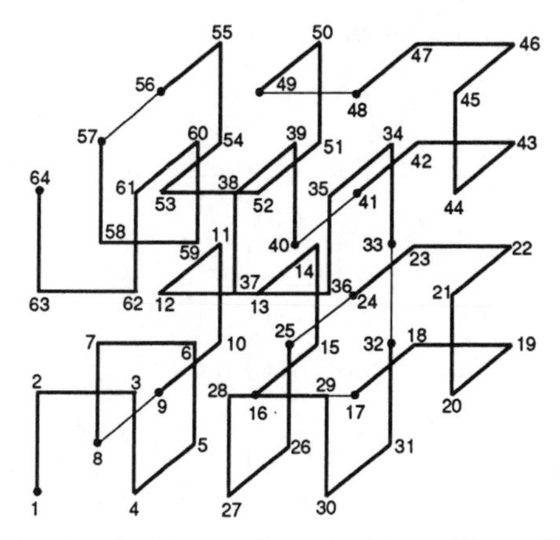

Fig. 2.8.2. Second Step in the Generation of a Three-Dimensional Hilbert Curve

By construction, the image of t lies in the $(\omega_1 + 1)$-th subcube of the first partition, the $(\omega_2 + 1)$-th subcube of the second partition of the $(\omega_1 + 1)$-th subcube of the first partition, etc.... The subcubes of the first partition, oriented as indicated by the clamp-like polygons in Fig. 2.8.2, may be obtained from $\mathcal{W}$ by similarity transformations. The subcubes of the second partition are then obtained by an application of these similarity transformations to the already partitioned cube $\mathcal{W}$, etc.... .

We call the similarity transformation that maps $\mathcal{W}$ onto the first subcube of the first partition $\mathfrak{H}_0$, the transformation that maps $\mathcal{W}$ onto the second subcube $\mathfrak{H}_1$, etc.... We denote the image of t by $f_h^{(3)}(t)$ and obtain, in view of

$$0_{\dot{8}}\omega_1\omega_2\omega_3\ldots\omega_n = 0_{\dot{8}}\omega_1\omega_2\omega_3\ldots\omega_n 000\ldots,$$

that

$$f_h^{(3)}(0_{\dot{8}}\omega_1\omega_2\omega_3\ldots\omega_n) = \mathfrak{H}_{\omega_1}\mathfrak{H}_{\omega_2}\mathfrak{H}_{\omega_3}\ldots\mathfrak{H}_{\omega_n}\mathfrak{H}_0\mathfrak{H}_0\mathfrak{H}_0\ldots\mathcal{W}. \tag{2.8.1}$$

To obtain the first (oriented) subcube in Fig. 2.8.2 from $\mathcal{W}$, we have to transform the clamp-like polygon in Fig. 2.8.1 into the one in the lower-left front corner of Fig. 2.8.2. This is accomplished by a reduction of the dimensions of $\mathcal{W}$ in the ratio 2:1, a rotation through 90° about the ξ-axis, and a reflection with respect to the (ξ, ζ) plane:

$$\frac{1}{2}\begin{pmatrix} 1 & 0 & 0 \\ 0 & -1 & 0 \\ 0 & 0 & 1 \end{pmatrix}\begin{pmatrix} 1 & 0 & 0 \\ 0 & 0 & -1 \\ 0 & 1 & 0 \end{pmatrix}\begin{pmatrix} \xi \\ \eta \\ \zeta \end{pmatrix} = \frac{1}{2}\begin{pmatrix} 1 & 0 & 0 \\ 0 & 0 & 1 \\ 0 & 1 & 0 \end{pmatrix}\begin{pmatrix} \xi \\ \eta \\ \zeta \end{pmatrix}.$$

To obtain the second subcube, we reduce the dimensions of $\mathcal{W}$ as before, rotate about the η-axis through 90°, reflect with respect to the (η, ζ)-plane, and shift the result by $\frac{1}{2}$ unit in the direction of the η-axis:

$$\frac{1}{2}\begin{pmatrix}-1&0&0\\0&1&0\\0&0&1\end{pmatrix}\begin{pmatrix}0&0&-1\\0&1&0\\1&0&0\end{pmatrix}\begin{pmatrix}\xi\\\eta\\\zeta\end{pmatrix}+\frac{1}{2}\begin{pmatrix}0\\1\\0\end{pmatrix}$$
$$=\frac{1}{2}\begin{pmatrix}0&0&1\\0&1&0\\1&0&0\end{pmatrix}\begin{pmatrix}\xi\\\eta\\\zeta\end{pmatrix}+\frac{1}{2}\begin{pmatrix}0\\1\\0\end{pmatrix}.$$

Continuing in this manner, we obtain eventually

$$\mathfrak{H}_0\begin{pmatrix}\xi\\\eta\\\zeta\end{pmatrix}=\frac{1}{2}\begin{pmatrix}1&0&0\\0&0&1\\0&1&0\end{pmatrix}\begin{pmatrix}\xi\\\eta\\\zeta\end{pmatrix}+\frac{1}{2}\begin{pmatrix}0\\0\\0\end{pmatrix}\triangleq\frac{1}{2}H_0x+\frac{1}{2}h_0$$
$$\mathfrak{H}_1\begin{pmatrix}\xi\\\eta\\\zeta\end{pmatrix}=\frac{1}{2}\begin{pmatrix}0&0&1\\0&1&0\\1&0&0\end{pmatrix}\begin{pmatrix}\xi\\\eta\\\zeta\end{pmatrix}+\frac{1}{2}\begin{pmatrix}0\\1\\0\end{pmatrix}\triangleq\frac{1}{2}H_1x+\frac{1}{2}h_1$$
$$\mathfrak{H}_2\begin{pmatrix}\xi\\\eta\\\zeta\end{pmatrix}=\frac{1}{2}\begin{pmatrix}1&0&0\\0&1&0\\0&0&1\end{pmatrix}\begin{pmatrix}\xi\\\eta\\\zeta\end{pmatrix}+\frac{1}{2}\begin{pmatrix}1\\1\\0\end{pmatrix}\triangleq\frac{1}{2}H_2x+\frac{1}{2}h_2$$
$$\mathfrak{H}_3\begin{pmatrix}\xi\\\eta\\\zeta\end{pmatrix}=\frac{1}{2}\begin{pmatrix}0&0&1\\-1&0&0\\0&-1&0\end{pmatrix}\begin{pmatrix}\xi\\\eta\\\zeta\end{pmatrix}+\frac{1}{2}\begin{pmatrix}1\\1\\1\end{pmatrix}\triangleq\frac{1}{2}H_3x+\frac{1}{2}h_3$$
$$\mathfrak{H}_4\begin{pmatrix}\xi\\\eta\\\zeta\end{pmatrix}=\frac{1}{2}\begin{pmatrix}0&0&-1\\-1&0&0\\0&1&0\end{pmatrix}\begin{pmatrix}\xi\\\eta\\\zeta\end{pmatrix}+\frac{1}{2}\begin{pmatrix}2\\1\\1\end{pmatrix}\triangleq\frac{1}{2}H_4x+\frac{1}{2}h_4$$
$$\mathfrak{H}_5\begin{pmatrix}\xi\\\eta\\\zeta\end{pmatrix}=\frac{1}{2}\begin{pmatrix}1&0&0\\0&1&0\\0&0&1\end{pmatrix}\begin{pmatrix}\xi\\\eta\\\zeta\end{pmatrix}+\frac{1}{2}\begin{pmatrix}1\\1\\1\end{pmatrix}\triangleq\frac{1}{2}H_5x+\frac{1}{2}h_5$$
$$\mathfrak{H}_6\begin{pmatrix}\xi\\\eta\\\zeta\end{pmatrix}=\frac{1}{2}\begin{pmatrix}0&0&-1\\0&1&0\\-1&0&0\end{pmatrix}\begin{pmatrix}\xi\\\eta\\\zeta\end{pmatrix}+\frac{1}{2}\begin{pmatrix}1\\1\\2\end{pmatrix}\triangleq\frac{1}{2}H_6x+\frac{1}{2}h_6$$
$$\mathfrak{H}_7\begin{pmatrix}\xi\\\eta\\\zeta\end{pmatrix}=\frac{1}{2}\begin{pmatrix}1&0&0\\0&0&-1\\0&-1&0\end{pmatrix}\begin{pmatrix}\xi\\\eta\\\zeta\end{pmatrix}+\frac{1}{2}\begin{pmatrix}0\\1\\2\end{pmatrix}\triangleq\frac{1}{2}H_7x+\frac{1}{2}h_7.$$

(We will see in Section 9.3 that $\mathcal{W}$ is the unique "fixed point" of the transformation $\mathfrak{H}_* : K(\mathbb{E}^3) \to K(\mathbb{E}^3)$, where $K(\mathbb{E}^3)$ is the space of all nonempty compact subsets of $\mathbb{E}^3$ that is defined by $\mathfrak{H}_*(\mathcal{A}) = \bigcup_{i=0}^{7} \mathfrak{H}_{i*}(\mathcal{A})$, $\mathcal{A} \in K(\mathbb{E}^3)$.)

Since

$$\mathfrak{H}_0\mathfrak{H}_0\mathfrak{H}_0\ldots\mathcal{W}=\lim_{n\to\infty}\mathfrak{H}_0^n\mathcal{W}=\lim_{n\to\infty}\frac{1}{2^n}H_0^n\mathcal{W}=\begin{pmatrix}0\\0\\0\end{pmatrix},$$

we obtain after substitution of H_j and h_j into (2.8.1) and with $H_{\omega_0} = I$ that

$$f_h^{(3)}(0_{\dot{8}}\omega_1\omega_2\omega_3\ldots\omega_n) = \sum_{j=1}^{n} \frac{1}{2^j} H_{\omega_0} H_{\omega_1} H_{\omega_2} \ldots H_{\omega_{j-1}} h_{\omega_j}$$

and, consequently,

$$f_h^{(3)}(0_{\dot{8}}\omega_1\omega_2\omega_3\ldots) = \sum_{j=1}^{\infty} \frac{1}{2^j} H_{\omega_0} H_{\omega_1} H_{\omega_2} \ldots H_{\omega_{j-1}} h_{\omega_j}. \tag{2.8.2}$$

Even though $H_2 = H_5 = H_0^2 = H_1^2 = H_6^2 = H_7^2 = H_3^3 = H_4^3 = I$, no further simplification (such as those shown in Section 2.4) is feasible, because the matrix products H_iH_j are, in general, not commutative. Therefore we cannot collect factors with the same subscript, as we have done in Section 2.4. Still, (2.8.2) can be evaluated by computer, and, since such an evaluation only involves the addition, subtraction, and multiplication of integers, the exact coordinates of the nodal points are obtained as fractions with 2^n as the denominator, as long as the computer's precision holds out. An alternate method for obtaining a three-dimensional Hilbert curve (and higher-dimensional Hilbert curves, for that matter) will be discussed in Section 6.9. See also Butz [1].

2.9. Problems

1. Show that $h_q = h_{q-1} + \frac{1}{2}H_{q-1}(2)$ for q=1,2,3.
2. Evaluate $f_h(0_{\dot{4}}203)$ and $f_h(0_{\dot{4}}10321)$ by formula (2.3.9).
3. Show that $\begin{pmatrix} \mathcal{R}e \\ \mathcal{I}m \end{pmatrix} (-1)^{e_{3j}} \mathrm{sgn}(q_j)\sqrt{(q_j-1)^2+1)}$
   $\times \exp[i\pi d_j/2 + i(-1)^{d_j} \arcsin(1/\sqrt{(q_j-1)^2+1})]$
   $= (-1)^{e_{0j}} \mathrm{sgn}(q_j) \begin{pmatrix} (1-d_j)q_j - 1 \\ 1 - d_jq_j \end{pmatrix}$, where e_{kj} is defined in (2.3.9a) and d_j defined in(2.3.12).
4. Use (2.4.3) to find $f_h(0_{\dot{4}}03021)$, $f_h(0_{\dot{4}}00001)$, $f_h(0_{\dot{4}}33333)$.
5. Let $t = 0_{\dot{4}}1023032$. Evaluate $\sum_{j=5}^{\infty}(1/2^j)(-1)^{e_j}\mathrm{sgn}(q_j)\begin{pmatrix} (1-d_j)q_j - 1 \\ 1 - d_jq_j \end{pmatrix}$, where d_j is defined in (2.3.12), and compare your result with $f_h(0.032)$ or $g_h(0.032)$, where g_h is obtained from f_h by an interchange of $f's$ components. Do the same for $t = 0_{\dot{4}}3001201$.
6. Verify the similarity transformations $\mathfrak{H}_0$ to $\mathfrak{H}_7$ shown in Section 2.8.
7. Write a computer program for the evaluation of (2.8.2) and find $f_h^{(3)}(0_{\dot{8}}000000001)$, $f_h^{(3)}(0_{\dot{8}}01234567$, and $f_h^{(3)}(0_{\dot{8}}777777777)$.
8. Explain why every point in $\mathcal{Q}$ has, under the mapping f_h, at most four preimages in $\mathcal{I}$.

9. Find a continuous mapping f from $(-\infty, \infty)$ onto the strip $0 \leq x \leq 1$, $-\infty < y < \infty$, so that $f(0) = \begin{pmatrix} 0 \\ 0 \end{pmatrix}$.
(For sundry applications of the parametrization of two- or higher-dimensional regions by means of space-filling curves, see Abend, Harley, and Kanal [1], Butz [1], [2], [3], Patrick, Anderson, and Bechtel [1], Bially [1], Bartholdi and Platzman [1], [2], [3], [4], and Bertsimas and Grigni [1].)

10. With $f_h = \begin{pmatrix} \varphi_h \\ \psi_h \end{pmatrix}$, find $\int_0^1 \varphi_h(t)\, dt$ and $\int_0^1 \psi_h(t)\, dt$.

12. Construct a space-filling curve by dissecting the trapezoid in Fig. 2.9 into four congruent replicas as indicated. (This problem is taken from Mioduszewski [2]. See also Golomb [1].)

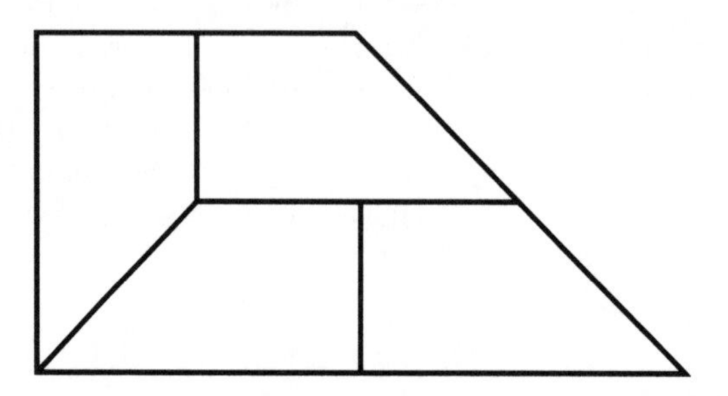

Fig. 2.9. Dissection of a Trapezoid

13. Show that the coordinate functions φ_h, ψ_h of the Hilbert curve satisfy the following functional equations:

$$\begin{aligned}
\varphi_h(t/4) &= (1/2)\psi_h(t) \\
\psi_h(t/4) &= (1/2)\varphi_h(t) \\
\varphi_h((1+t)/4) &= (1/2)\varphi_h(t) \\
\psi_h((1+t)/4)) &= 1/2 + (1/2)\psi_h(t)) \\
\varphi_h((2+t)/4) &= 1/2 + (1/2)\varphi_h(t) \\
\psi_h((2+t)/4) &= 1/2 + (1/2)\psi_h(t) \\
\varphi_h((3+t)/4) &= 1 - (1/2)\psi_h(t) \\
\psi_h((3+t)/4) &= 1/2 - (1/2)\varphi_h(t) \\
\varphi_h(t) + \varphi_h(1-t) &= 1 \\
\psi_h(t) &= \psi_h(1-t).
\end{aligned}$$

Chapter 3

Peano's Space-Filling Curve

3.1. Definition of Peano's Space-Filling Curve

Giuseppe Peano (1858–1932) was born in Spinetta, Italy, and died in Turin. He completed his studies at the University of Turin in 1880 and became a professor there in 1890. He held that position until his death. He is known for his pioneering work in symbolic logic, the axiomatic method, and for his contributions to mathematical analysis.

It was Peano who introduced the symbol "$\in$" (because it is the initial of the Greek singular copula '$\epsilon\sigma\tau$) for set membership and "$\ni$" for "such that." B. Russel once said that until he got hold of Peano, it never struck him that symbolic logic would be any use for the principles of mathematics because he knew the Boolean stuff and found it useless. Peano's best known contribution to analysis is probably his proof of the existence of a solution of a system of first-order differential equations (without Lipschitz condition). In his "Arithmetices principia," he laid down the foundations for the arithmetic of natural numbers as explicit axioms, expressed in arithmetical and logical symbols. We still speak of "Peano's axioms" even though Peano himself attributes them to Dedekind. Ivor Grattan-Guiness [1, p. 27] appraises Peano by pointing out his remarkable capacity to spot counter-examples, to point out distinctions in set theory that others overlooked, and to feel the utility of linguistic analysis in the clarification of concepts. He was not, however, a systematic conceptualist of large-scale mathematical theories or a constructor of coherent philosophical edifices. (For more details, see Kennedy [1], Orman Quine [1], and Grattan-Guiness [1].)

Peano [1], defined a map f_p from $\mathcal{I}$ to $\mathcal{Q}$ in terms of the operator

$$kt_j = 2 - t_j (t_j = 0, 1, 2)$$

as follows:

$$f_p(0_{\dot{3}}t_1t_2t_3t_4\ldots) = \begin{pmatrix} 0_{\dot{3}}t_1(k^{t_2}t_3)(k^{t_2+t_4}t_5)\ldots \\ 0_{\dot{3}}(k^{t_1}t_2)(k^{t_1+t_3}t_4)\ldots \end{pmatrix}, \tag{3.1.1}$$

where k^ν denotes the νth iterate of k, and demonstrated that it is surjective and continuous.

Since every finite ternary may also be written as an infinite ternary with infinitely many trailing 3's,

$$0_{\dot{3}}t_1t_2t_3\ldots t_nt = 0_{\dot{3}}t_1t_2t_3\ldots t_n(t-1)\overline{2}, \quad t = 1 \text{ or } 2,$$

we have to make sure that the value of f_p is independent of the representation. We assume, for the time being, that $n = 2m$ and obtain for the first component φ_p of f_p from (3.1.1), using the abbreviating notation $\tau = t_2 + \cdots + t_{2m}$, that

$$\begin{aligned}\varphi_p(0_{\dot{3}}t_1t_2t_3\ldots t_{2m}t) &= 0_{\dot{3}}t_1(k^{t_2}t_3)(k^{t_2+t_4}t_5)\ldots(k^\tau t)(\overline{k^\tau 0}) \\ &= \begin{cases} 0_{\dot{3}}t_1(k^{t_2}t_3)(k^{t_2+t_4}t_5)\ldots t \text{ if } \tau \text{ is even} \\ 0_{\dot{3}}t_1(k^{t_2}t_3)\ldots(k^{t_2\cdots+t_{2m-2}}t_{2m-1})(3-t) \text{ if } \tau \text{ is odd}\end{cases}\end{aligned}$$

and

$$\varphi_p(0_{\dot{3}}t_1t_2t_3\ldots t_{2m}(t-1)\overline{2}) = 0_{\dot{3}}t_1(k^{t_2}t_3)\ldots(k^\tau(t-1))(k^{\tau+2}2)(k^{\tau+4}2)\ldots$$

$$= \begin{cases} 0_{\dot{3}}t_1(k^{t_2}t_3)(k^{t_2+t_4}t_5)\ldots(t-1)\overline{2} = 0.t_1(k^{t_2}t_3)\ldots t \text{ if } \tau \text{ is even} \\ 0_{\dot{3}}t_1(k^{t_2}t_3)(k^{t_2+t_4}t_5)\ldots(2-(t-1)) = 0_{\dot{3}}t_1(k^{t_2}t_3)(k^{t_2+t_4}t_5)\ldots(3-t), \\ \text{if } \tau \text{ is odd.}\end{cases}$$

We leave it to the reader to deal with the case $n = 2m+1$ and the second component ψ_p of f_p, and proceed with the understanding that we obtain the same value for f_p either way.

The map f_p as defined in (3.1.1) is surjective: Let

$$(0_{\dot{3}}\beta_1\beta_2\beta_3\beta_4\ldots, 0_{\dot{3}}\gamma_1\gamma_2\gamma_3\gamma_4\ldots) \in \mathcal{Q}.$$

We have to show that there is a $t = 0_{\dot{3}}t_1t_2t_3t_4\ldots \in \mathcal{I}$ such that

$$f_p(t) = \begin{pmatrix} 0_{\dot{3}}\beta_1\beta_2\beta_3\beta_4\ldots \\ 0_{\dot{3}}\gamma_1\gamma_2\gamma_3\gamma_4\ldots \end{pmatrix}.$$

Comparing corresponding ternary places, we obtain with $t_0 = 0$

$$\beta_n = k^{t_0+t_2+t_4+\cdots+t_{2n-2}}t_{2n-1}, \quad \gamma_n = k^{t_1+t_3+\cdots+t_{2n-1}}t_{2n}.$$

Noting that $k \circ k(t_j) = 2-(2-t_j) = t_j$, i.e., k is its own inverse, we obtain from the above that

$$t_{2n-1} = k^{t_0+t_2+t_4+\cdots+t_{2n-2}}\beta_n, t_{2n} = k^{t_1+t_3+\cdots+t_{2n-1}}\gamma_n,$$

which we can solve successively for $t_1, t_2, t_3, \ldots$.

The map f_p, as defined in (3.1.1), is continuous:

The following proof, which is the "most convincing" proof of the continuity of the Paeno curve that we have come across, was sent to us by Federico Prat-Villar from the Universidad Polytécnica de Valencia, Spain.

We will first show that φ_p is continuous from the right for all $t \in [0,1)$. Let

$$t_0 = 0_{\dot{3}}t_1t_2t_3\ldots t_{2n}t_{2n+1}\cdots$$

be that representation of t_0 that does not have infinitely many trailing $2'$s and let

$$\delta = 1/3^{2n} - 0_{\dot{3}}000\ldots t_{2n+1}t_{2n+2}\cdots .$$

Since

$$\begin{aligned} t_0 + \delta &= 0_{\dot{3}}t_1t_2t_3\ldots t_{2n}t_{2n+1}\cdots + 1/3^{3n} - 0_{\dot{3}}000\ldots 0t_{2n+1}t_{2n+2}\cdots \\ &= 0_{\dot{3}}t_1t_2t_3\ldots t_{2n}\bar{2}, \end{aligned}$$

any $t \in [t_0, t_0+\delta)$ has to agree with t_0 in the first $2n$ digits after the ternary point:

$$t = 0_{\dot{3}}t_1t_2t_3\ldots t_{2n}\tau_{2n+1}\tau_{2n+2}\cdots .$$

With $\epsilon = t_2 + t_4 + t_6 + \cdots + t_{2n}$,

$$\begin{aligned} |\varphi_p(t) - \varphi_p(t_0)| &= |0_{\dot{3}}t_1(k^{t_2}t_3)\ldots(k^{\epsilon}\tau_{2n+1})\cdots \\ &\qquad - 0_{\dot{3}}t_1(k^{t_2}t_3)\ldots(k^{\epsilon}t_{2n+1})\ldots| \\ &\leq |k^{\epsilon}\tau_{2n+1} - k^{\epsilon}t_{2n+1}|/3^{n+1} + |k^{\epsilon+\tau_{2n+2}}\tau_{2n+3} \\ &\quad - k^{\epsilon+t_{2n+2}}t_{2n+3}|/3^{n+2} + \cdots \\ &\leq (2/3^{n+1})(1 + 1/3 + 1/9 + \cdots) = 1/3^n \to 0 \quad \text{as} \quad n \to \infty. \end{aligned}$$

To show that φ_p is continuous from the left in (0,1], we pick that ternary representation of t_0 which is not finite and let $\delta = 0_{\dot{3}}000\ldots 0t_{2n+1}t_{2n+2}\cdots$. Then, $t_0 - \delta = 0_{\dot{3}}t_1t_2t_3\ldots t_{2n}$ and we see as before that for t to lie in $(t_0 - \delta, t_0]$, its ternary representation has to agree with the ternary representation of t_0 in the first $2n$ digits after the ternary point. We now proceed as before. Since φ_p is continuous from the right in $[0,1)$ and continuous from the left in $(0,1]$, it is continuous in $[0,1]$. The continuity of ψ_p follows from $\psi_p(t) = 3\varphi_p(t/3)$.

We summarize our results in the following:

(3.1) Theorem. *$f_p : \mathcal{I} \to \mathcal{Q}$, as defined in (3.1.1), represents a space-filling curve. We call it the Peano curve.*

We will see in Section 3.3 how this curve may be generated by Hilbert's geometric principle.

3.2. Nowhere Differentiability of the Peano Curve

It is noteworthy that Peano [1] finds it necessary to give an elaborate explanation of the ternary representation of a number, while dispatching the question as to the differentiability of his curve with just one sentence, the last sentence in his paper: "*Ces x et y*—we called them φ_p and ψ_p—*fonctions continues de la variable t, manquent toujours de dérivée.*" ("These x and y, continuous functions of the variable t, lack a derivative altogether.") It was not until ten years later that Moore published a proof of the nowhere differentiability of the Peano curve in Moore [1]. We won't ever know how Peano convinced himself of the nowhere differentiability of his curve, but his proof could not possibly have been any simpler than the one we offered in Sagan [3], and which we will reproduce here.

(3.2) Theorem. *The Peano curve, as defined by (3.1.1), is nowhere differentiable.*

Proof. For any $t = 0_{\dot{3}}t_1t_2t_3\ldots t_{2n}t_{2n+1}t_{2n+2}\cdots \in [0,1]$, we define $t_n = 0_{\dot{3}}t_1t_2t_3\ldots t_{2n}\tau_{2n+1}t_{2n+2}\cdots$, where $\tau_{2n+1} = t_{2n+1} + 1 \pmod 2$. Then, $|t-t_n| = 1/3^{2n+1}$. By (3.1.1), $\varphi_p(t)$ and $\varphi_p(t_n)$ only differ in the $(n+1)$-th ternary place and we have

$$|\varphi_p(t) - \varphi_p(t_n)| = |k^{t_2+\cdots+t_{2n}}t_{2n+1} - k^{t_2+\cdots+t_{2n}}\tau_{2n+1}|/3^{n+1} = 1/3^{n+1}$$

and hence,

$$|\varphi_p(t) - \varphi_p(t_n)|/|t-t_n| = 3^n \to \infty.$$

The nowhere differentiability of ψ_p follows from $\psi_p(t) = 3\varphi_p(t/3)$. □

3.3. Geometric Generation of the Peano Curve

Although there is no hint about a geometric interpretation to be found anywhere in Peano [1], Hilbert was undoubtedly led to his generating principle by the following consideration: By (3.1.1),

$$f_p(0_{\dot{3}}00t_3t_4t_5\ldots) = \begin{pmatrix} 0_{\dot{3}}0\xi_2\xi_3\xi_4\cdots \\ 0_{\dot{3}}0\eta_2\eta_3\eta_4\ldots \end{pmatrix},$$

meaning that the interval $[1, \frac{1}{9}]$ is mapped into the subsquare no.1 in Fig. 3.3.1. By the same token,

$$f_p(0_{\dot{3}}01t_3t_4t_5\ldots) = \begin{pmatrix} 0_{\dot{3}}0\xi_2'\xi_3'\xi_4'\cdots \\ 0_{\dot{3}}1\eta_2'\eta_3'\eta_4'\ldots \end{pmatrix},$$

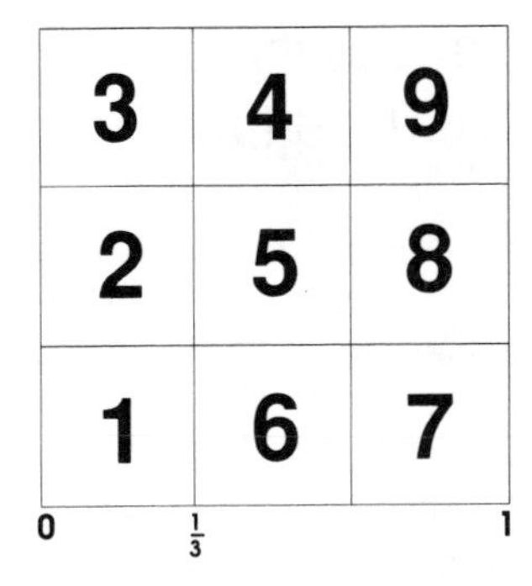

Fig. 3.3.1. Peano's Mapping

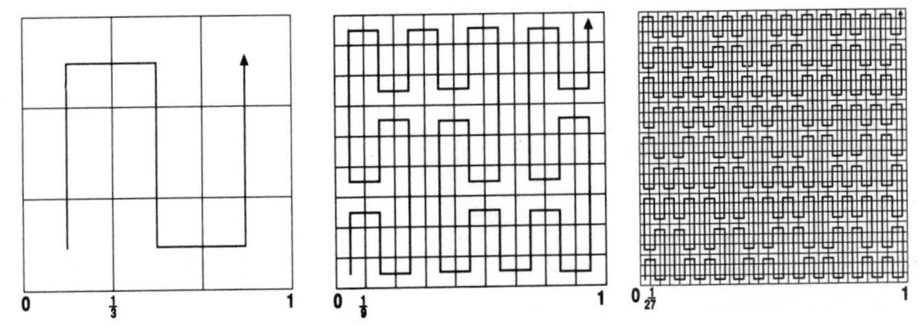

Fig. 3.3.2. Geometric Generation of the Peano Curve

and we see that the interval $[\frac{1}{9}, \frac{2}{9}]$ is mapped into subsquare no. 2 in Fig. 3.3.1. We continue in this manner and recognize that the interval $[\frac{j-1}{9}, \frac{j}{9}], j = 1, 2, 3, \ldots, 9$, is mapped into subsquare no. j in Fig. 3.3.1.

This suggests that we may obtain Peano's curve if we partitioned $\mathcal{I}$ into 3^{2n} congruent subintervals and mapped them into the $3^{2n}, n = 1, 2, 3, \ldots$ subsquares in the order that we have indicated in Fig. 3.3.2 for the first three steps.

Our conditions for the generation of a space-filling curve appear to be satisfied: adjacent intervals are mapped onto adjacent squares with an edge in common, and each mapping preserves the preceding one. We can, therefore, define a mapping $g_p : \mathcal{I} \to \mathcal{Q}$ as in Section 2.1 and show, just as we did then, that the mapping is surjective and continuous, i.e., it represents a space-filling curve. In view of the construction, we suspect that it represents the Peano curve and Moore has shown just that (Moore [1]). Let us call it the *geometric* Peano curve for the time being. Moore's proof covers the more general case where $\mathcal{I}$ and $\mathcal{Q}$ are dissected into $(2m + 1)^2$ replicas at each step. In order to avoid Moore's complicated recursion formulas, we will instead use the techniques of Sections 2.3 and 2.4 to represent g_p algebraically and then show that it is identical with f_p as defined in (3.1.1).

3.4. Proof that the Peano Curve and the Geometric Peano Curve are the Same

We observe that the space-filling curve $g_{p*}(\mathcal{I})$, obtained by the geometric generating process in the preceding section, emanates from $(0,0)$ and terminates diagonally across at $(1,1)$. The subsquares will have to be oriented in such a manner that the exit point from one subsquare coincides with the entry point of the next subsquare. We have illustrated this in Fig. 3.4.1. The bold arrows indicate the direction from entry point to exit point. As in Section 2.3, we use complex representation to obtain an algebraic representation of the required transformations. We see immediately that subsquares nos. 1, 3, 7, and 9 are obtained from $\mathcal{Q}$ by shrinking $\mathcal{Q}$ uniformly towards the origin at the ratio 3:1 and then translating the shrunken subsquare by the required amounts to obtain

$$\mathfrak{P}_0 z = \frac{1}{3}z, \mathfrak{P}_2 z = \frac{1}{3}z + \frac{2i}{3}, \mathfrak{P}_6 z = \frac{1}{3}z + \frac{2}{3}, \mathfrak{P}_8 z = \frac{1}{3}z + \frac{2}{3} + \frac{2i}{3}.$$

To obtain subsquare 2, we have to shrink, reflect on the imaginary axis, and translate:

$$\mathfrak{P}_1 z = -\frac{1}{3}\overline{z} + \frac{1}{3} + \frac{i}{3}.$$

In order to obtain subsquare no. 4, we have to shrink, reflect on the real axis, and translate:

$$\mathfrak{P}_3 z = \frac{1}{3}\overline{z} + i + \frac{1}{3}.$$

We obtain subsquare no. 5 by shrinking, rotating through 180°, and translating:

$$\mathfrak{P}_4 z = -\frac{1}{3}z + \frac{2}{3} + \frac{2i}{3}.$$

Subsquares no. 6 and no. 8 are obtained as subsquares no. 4 and no. 2, respectively, except for the translations:

$$\mathfrak{P}_5 z = \frac{1}{3}\overline{z} + \frac{1}{3} + \frac{i}{3}, \quad \mathfrak{P}_7 = -\frac{1}{3}\overline{z} + 1 + \frac{i}{3}.$$

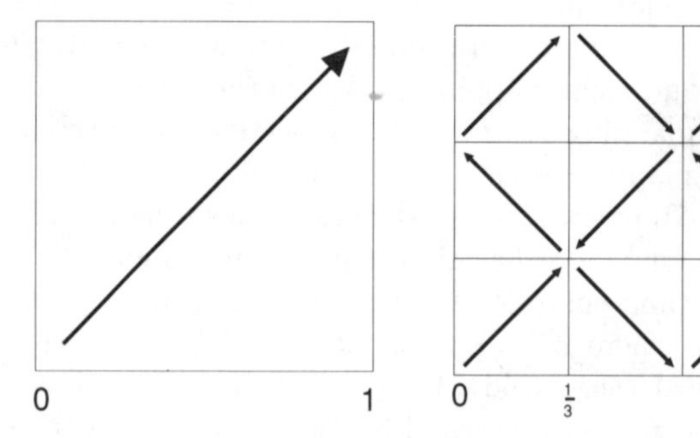

Fig. 3.4.1. Mapping the Square

(As in Section 2.3, we have enumerated the transformations so that their index matches up with the first digit in t, were t represented in *eneadic* form, rather than with the subsquare number.) Decomposition of these transformations into real and imaginary parts yields the following similarity transformations:

$$\begin{aligned}
\mathfrak{P}_0\begin{pmatrix}\xi\\ \eta\end{pmatrix} &= \frac{1}{3}\begin{pmatrix}1 & 0\\ 0 & 1\end{pmatrix}\begin{pmatrix}\xi\\ \eta\end{pmatrix} + \frac{1}{3}\begin{pmatrix}0\\ 0\end{pmatrix} \triangleq \frac{1}{3}P_0x + \frac{1}{3}p_0\\
\mathfrak{P}_1\begin{pmatrix}\xi\\ \eta\end{pmatrix} &= \frac{1}{3}\begin{pmatrix}-1 & 0\\ 0 & 1\end{pmatrix}\begin{pmatrix}\xi\\ \eta\end{pmatrix} + \frac{1}{3}\begin{pmatrix}1\\ 1\end{pmatrix} \triangleq \frac{1}{3}P_1x + \frac{1}{3}p_1\\
\mathfrak{P}_2\begin{pmatrix}\xi\\ \eta\end{pmatrix} &= \frac{1}{3}\begin{pmatrix}1 & 0\\ 0 & 1\end{pmatrix}\begin{pmatrix}\xi\\ \eta\end{pmatrix} + \frac{1}{3}\begin{pmatrix}0\\ 2\end{pmatrix} \triangleq \frac{1}{3}P_2x + \frac{1}{3}p_2\\
\mathfrak{P}_3\begin{pmatrix}\xi\\ \eta\end{pmatrix} &= \frac{1}{3}\begin{pmatrix}1 & 0\\ 0 & -1\end{pmatrix}\begin{pmatrix}\xi\\ \eta\end{pmatrix} + \frac{1}{3}\begin{pmatrix}1\\ 3\end{pmatrix} \triangleq \frac{1}{3}P_3x + \frac{1}{3}p_3\\
\mathfrak{P}_4\begin{pmatrix}\xi\\ \eta\end{pmatrix} &= \frac{1}{3}\begin{pmatrix}-1 & 0\\ 0 & -1\end{pmatrix}\begin{pmatrix}\xi\\ \eta\end{pmatrix} + \frac{1}{3}\begin{pmatrix}2\\ 2\end{pmatrix} \triangleq \frac{1}{3}P_4x + \frac{1}{3}p_4\\
\mathfrak{P}_5\begin{pmatrix}\xi\\ \eta\end{pmatrix} &= \frac{1}{3}\begin{pmatrix}1 & 0\\ 0 & -1\end{pmatrix}\begin{pmatrix}\xi\\ \eta\end{pmatrix} + \frac{1}{3}\begin{pmatrix}1\\ 1\end{pmatrix} \triangleq \frac{1}{3}P_5x + \frac{1}{3}p_5\\
\mathfrak{P}_6\begin{pmatrix}\xi\\ \eta\end{pmatrix} &= \frac{1}{3}\begin{pmatrix}1 & 0\\ 0 & 1\end{pmatrix}\begin{pmatrix}\xi\\ \eta\end{pmatrix} + \frac{1}{3}\begin{pmatrix}2\\ 0\end{pmatrix} \triangleq \frac{1}{3}P_6x + \frac{1}{3}p_6\\
\mathfrak{P}_7\begin{pmatrix}\xi\\ \eta\end{pmatrix} &= \frac{1}{3}\begin{pmatrix}-1 & 0\\ 0 & 1\end{pmatrix}\begin{pmatrix}\xi\\ \eta\end{pmatrix} + \frac{1}{3}\begin{pmatrix}3\\ 1\end{pmatrix} \triangleq \frac{1}{3}P_7x + \frac{1}{3}p_7\\
\mathfrak{P}_8\begin{pmatrix}\xi\\ \eta\end{pmatrix} &= \frac{1}{3}\begin{pmatrix}1 & 0\\ 0 & 1\end{pmatrix}\begin{pmatrix}\xi\\ \eta\end{pmatrix} + \frac{1}{3}\begin{pmatrix}2\\ 2\end{pmatrix} \triangleq \frac{1}{3}P_8x + \frac{1}{3}p_8.
\end{aligned} \tag{3.4.1}$$

(The transformation $\mathfrak{P}_* : K(\mathbb{E}^2) \to K(\mathbb{E}^2)$, where $K(\mathbb{E}^2)$ is the space of all compact non-empty subsets of $\mathbb{E}^2$, which is defined by $\mathfrak{P}_*(\mathcal{A}) = \bigcup_{i=0}^{8} \mathfrak{P}_{i*}(\mathcal{A})$, $\mathcal{A} \in K(\mathbb{E}^2)$, has the unique "fixed point" $\mathcal{Q}$, as we will see in Section 9.3.)

We will now use these transformations to show:

(3.4) Theorem. *The Peano curve of Section 3.1 and the geometric Peano curve of Section 3.3 are identical:* $g_p(t) = f_p(t)$ *for all* $t \in [0, 1]$.

Proof. We will show that $g_p(t) = f_p(t)$ for all $t \in \mathcal{D}$, where $\mathcal{D} = \{0_{\dot{3}}t_1t_2t_3 \ldots t_{2n} | t_j = 0, 1, \text{ or } 2, n = 1, 2, 3, \ldots\}$ is dense in $\mathcal{I}$, and we will then be able to conclude that $f_p(t) = g_p(t)$ for all $t \in I$.

We note that

$$0_{\dot{3}}t_1t_2 \ldots t_{2n-1}t_{2n} \ldots = 0_{\dot{9}}(3t_1 + t_2)(3t_3 + t_4) \ldots (3t_{2n-1} + t_{2n}) \ldots$$

and we obtain, as in Section 2.3,

$$g_p(t) = \lim_{n\to\infty} \mathfrak{P}_{3t_1+t_2}\mathfrak{P}_{3t_3+t_4} \cdots \mathfrak{P}_{3t_{2n-1}+t_{2n}}\mathcal{Q}$$

and, in particular,

$$g_p(0_{\dot{3}}t_1t_2\ldots t_{2n-1}t_{2n}) = \mathfrak{P}_{3t_1+t_2}\mathfrak{P}_{3t_3+t_4}\cdots\mathfrak{P}_{3t_{2n-1}+t_{2n}}\begin{pmatrix}0\\0\end{pmatrix}. \qquad (3.4.2)$$

We will show by induction with respect to n that

$$f_p(0_{\dot{3}}t_1t_2\ldots t_{2n-1}t_{2n}) = g_p(0_{\dot{3}}t_1t_2\ldots t_{2n-1}t_{2n}) \qquad (3.4.3)$$

for all n.

To show that (3.4.3) is true for $n = 1$, we have to check all nine possibilities $3t_1 + t_2 = 0, 1, 2, 3, \ldots, 8$. We will only check two, and leave the others to the reader.

First, we consider $t = 0_{\dot{3}}00$, and we have from (3.4.2) that

$$g_p(0_{\dot{3}}00) = \mathfrak{P}_0\begin{pmatrix}0\\0\end{pmatrix} = \begin{pmatrix}0\\0\end{pmatrix}$$

and from (3.1.1)

$$f_p(0_{\dot{3}}00) = \begin{pmatrix}0_{\dot{3}}0(k^00)(k^00)\ldots\\0_{\dot{3}}(k^00)(k^00)\ldots\end{pmatrix} = \begin{pmatrix}0\\0\end{pmatrix}.$$

Next, we consider $t = 0_{\dot{3}}12$, and note that $3t_1 + t_2 = 5$. Hence,

$$g_p(0_{\dot{3}}12) = \mathfrak{P}_5\begin{pmatrix}0\\0\end{pmatrix} = \begin{pmatrix}\frac{1}{3}\\\frac{1}{3}\end{pmatrix},$$

while

$$f_p(0_{\dot{3}}12) = \begin{pmatrix}0_{\dot{3}}1(k^00)(k^20)\ldots\\0_{\dot{3}}(k^12)(k^10)(k^10)\ldots\end{pmatrix} = \begin{pmatrix}0_{\dot{3}}1\\0_{\dot{3}}0\overline{2}\end{pmatrix} = \begin{pmatrix}\frac{1}{3}\\\frac{1}{3}\end{pmatrix}.$$

The other seven cases are similar.

We make the induction hypothesis that

$$f_p(0_{\dot{3}}t_3t_4t_5\ldots t_{2n}) = g_p(0_{\dot{3}}t_3t_4t_5\ldots t_{2n}). \qquad (3.4.4)$$

Let

$$f_p(0_{\dot{3}}t_3t_4t_5\ldots t_{2n}) = \begin{pmatrix}0_{\dot{3}}\xi_2\xi_3\xi_4\ldots\\0_{\dot{3}}\eta_2\eta_3\eta_4\ldots\end{pmatrix}, \qquad (3.4.5)$$

where

$$\xi_2 = t_3, \xi_3 = k^{t_4}t_5, \xi_4 = k^{t_4+t_6}t_7, \ldots$$

and

$$\eta_2 = k^{t_3}t_4, \eta_3 = k^{t_3+t_5}t_6, \eta_4 = k^{t_3+t_5+t_7}t_8, \ldots .$$

We have to show that (3.4.3) is true, given (3.4.5). Again, we have to check all nine cases for $3t_1 + t_2$. We pick 2 and leave the others to the reader:

First, take $t = 0_{\dot{3}}02$, i.e., $3t_1 + t_2 = 2$. In view of (3.4.4) and (3.4.5), we obtain

$$g_p(0_{\dot{3}}02t_3t_4\dots t_{2n}) = \mathfrak{P}_2 \begin{pmatrix} 0_{\dot{3}}\xi_2\xi_3\xi_4\dots \\ 0_{\dot{3}}\eta_2\eta_3\eta_4\dots \end{pmatrix} = \begin{pmatrix} 0_{\dot{3}}0\xi_2\xi_3\xi_4\dots \\ 0_{\dot{3}}0\eta_2\eta_3\eta_4\dots \end{pmatrix} + \begin{pmatrix} 0 \\ 0_{\dot{3}}2 \end{pmatrix}$$
$$= \begin{pmatrix} 0_{\dot{3}}0\xi_2\xi_3\xi_4\dots \\ 0_{\dot{3}}2\eta_2\eta_3\eta_4\dots \end{pmatrix},$$

while

$$f_p(0_{\dot{3}}02t_3t_4\dots t_{2n}) = \begin{pmatrix} 0_{\dot{3}}0(k^2t_3)(k^{2+t_4}t_5)\dots \\ 0_{\dot{3}}(k^0 2)(k^{0+t_3}t_4)\dots \end{pmatrix} = \begin{pmatrix} (0_{\dot{3}}0\xi_2\xi_3\xi_4\dots \\ 0_{\dot{3}}2\eta_2\eta_3\eta_4\dots \end{pmatrix}.$$

Next, let us look at $t = 0_{\dot{3}}21$ with $3t_1 + t_2 = 7$. We have

$$g_p(0_{\dot{3}}21t_3t_4t_5\dots t_{2n}) = \mathfrak{P}_7 \begin{pmatrix} 0_{\dot{3}}\xi_2\xi_3\xi_4\dots \\ 0_{\dot{3}}\eta_2\eta_3\eta_4\dots \end{pmatrix} = \begin{pmatrix} -0_{\dot{3}}0\xi_2\xi_3\xi_4\dots \\ 0_{\dot{3}}0\eta_2\eta_3\eta_4\dots \end{pmatrix} + \begin{pmatrix} 1 \\ \frac{1}{3} \end{pmatrix}$$
$$= \begin{pmatrix} 0_{\dot{3}}\overline{2} - 0_{\dot{3}}0\xi_2\xi_3\xi_4\dots \\ 0_{\dot{3}}1\eta_2\eta_3\eta_4\dots \end{pmatrix} = \begin{pmatrix} 0_{\dot{3}}2(2-\xi_2)(2-\xi_3)(2-\xi_4)\dots \\ 0_{\dot{3}}1\eta_2\eta_3\eta_4\dots \end{pmatrix}$$
$$= \begin{pmatrix} 0_{\dot{3}}2(k\xi_2)(k\xi_3)(k\xi_4)\dots \\ 0_{\dot{3}}1\eta_2\eta_3\eta_4\dots \end{pmatrix}.$$

On the other hand,

$$f_p(0_{\dot{3}}21t_3t_4t_5\dots t_{2n}) = \begin{pmatrix} 0_{\dot{3}}2(k^1t_3)(k^{1+t_4}t_5)\dots \\ 0_{\dot{3}}(k^2 1)(k^{2+t_3}t_4)\dots \end{pmatrix} = \begin{pmatrix} 0_{\dot{3}}2(k\xi_2)(k\xi_3)\dots \\ 0_{\dot{3}}1\eta_2\eta_3\eta_4\dots \end{pmatrix}.$$

We leave the remaining seven cases for the reader and consider (3.4.3) as proved. The polygonal lines in Fig. 3.3.2, with the beginning point joined to $(0,0)$ and the endpoint joined to $(1,1)$ by straight line segments, form a sequence of continuous curves that converges uniformly to g_p on $\mathcal{I}$. Hence, g_p is continuous on $\mathcal{I}$. We know already that f_p is continuous on $\mathcal{I}$ and we have just seen that

$$h(t) = f_p(t) - g_p(t) = 0 \quad \text{for all} \quad t \in \mathcal{D}.$$

Let $t \in \mathcal{I}\backslash\mathcal{D}$. There exists a sequence $\{t_k\} \to t$ with $t_k \in \mathcal{D}$ and, in view of the continuity of $h, h(t) = \lim_{k\to\infty} h(t_k) = \lim_{k\to\infty} 0 = 0$, i.e., $f_p(t) = g_p(t)$ for all $t \in \mathcal{I}$. □

We could, as in Section 2.4, use the similarity transformations (3.4.1) to obtain an analytic-arithmetic representation of f_p. Even though four of the nine P_j may be omitted, representing the identity matrix, and only three of the remaining five are distinct, the formula would not be as simple as (2.4.3) and not nearly as elegant as Peano's representation (3.1.1) and, hence, of little interest. This is particularly true in view of the fact that

(3.1.1) lends itself quite readily to an evaluation by computer. We have displayed a program in BASIC in Appendix A.1.2.

3.5. Cesàro's Representation of the Peano Curve

Cesàro gives the following analytic-arithmetic representation of the Peano curve as defined in (3.1.1):

$$\varphi(t)=\sum_{n=1}^{\infty} f_{2n-1}(t)/3^n,\ \psi(t)=\sum_{n=1}^{\infty} f_{2n}(t)/3^n, \tag{3.5.1}$$

where

$$f_m(t)=1+([3^m t]-3[3^{m-1}t]-1)(-1)^{[3t]+[3^2t]+\cdots+[3^{m-1}t]} \tag{3.5.1a}$$

(Cesàro [1]). Our formula (3.5.1a) differs from Cesàro's formula by 1 in order to adjust for the discrepancy caused by Peano's mapping onto $\mathcal{Q}=[0,1]^2$, while Cesàro maps onto $\left[-\frac{1}{2},\frac{1}{2}\right]^2$. Although $f_1(t)$ is not defined by (3.5.1a), we are sure that Cesàro meant it to be

$$f_1(t)=1+([3t]-3[3t]-1)(-1)^{[t]}. \tag{3.5.1b}$$

Moore ([1], p. 73, footnote †) notes that this formula is incorrect for $t=1$. Indeed, if one substitutes $t=1$ into (3.5.1), one obtains $\varphi(1)=1$ and $\psi(1)=0$ instead of 1 in both cases. Sierpiński, who deals with this matter, ignores the case $t=1$ entirely (Sierpiński [2], pp. 202–203). Apparently, this is where this matter was left to rest.

Cesàro's formula can be salvaged if one makes some formal changes in Peano's definition (3.1.1) and some substantive change in Cesàro's representation (3.1.1).

First, let us replace definition (3.1.1) by

$$\begin{aligned}\varphi_p(t_{0\dot{3}}t_1t_2t_3\ldots)&=0_{\dot{3}}(k^{t_0}t_1)(k^{t_0+t_2}t_3)(k^{t_0+t_2+t_4}t_5)\ldots\\ \psi_p(t_{0\dot{3}}t_1t_2t_3\ldots)&=t_{0\dot{3}}(k^{t_1}t_2)(k^{t_1+t_3}t_4)(k^{t_1+t_3+t_5}t_6)\ldots .\end{aligned} \tag{3.5.2}$$

If $t_{0\dot{3}}t_1t_2t_3\ldots\in[0,1)$, then $t_0=0$ and (3.5.2) reverts to (3.1.1). If $t_{0\dot{3}}t_1t_2t_3\ldots=1$, then $t_0=1$ and $t_j=0$ for all $j\geq 1$ and $\varphi_p(1)=0_{\dot{3}}(k0)(k0)(k0)\ldots=0.\overline{2}=1, \psi_p(1)=1_{\dot{3}}(k^00)(k^00)(k^00)\ldots=1$.

Next, we change Cesàro's representation to

$$\varphi_p(t)=\sum_{n=1}^{\infty} f_{2n-1}(t)/3^n,\ \psi_p(t)=[t]+\sum_{n=1}^{\infty} f_{2n}(t)/3^n \tag{3.5.3}$$

with

$$f_m(t)=1+([3^m t]-3[3^{m-1}t]-1)(-1)^{[t]+[3t]+[3^2t]+\cdots+[3^{m-1}t]}. \tag{3.5.3a}$$

(Note that we have added $[t]$ to the expression representing ψ_p in (3.5.3) and also to the exponent of (-1) in (3.5.3a).) In order to verify (3.5.3) and (3.5.3a), let us note that, with $t = 0_{\dot{3}}t_1t_2\ldots t_n\ldots \in [0,1)$, and with the understanding that every ternary with infinitely many trailing 2's is re-written as a finite ternary,

$$3^n t = t_1t_2\ldots t_{n\dot{3}}t_{n+1}t_{n+2}\ldots, [3^n t] = (t_1\ldots t_n)_3$$
$$[3^{n-1}t] = (t_1\ldots t_{n-1})_3, 3[3^{n-1}t] = (t_1\ldots t_{n-1}0)_3,$$

and, hence,

$$[3^n t] - 3[3^{n-1}t] = t_n \quad \text{for} \quad n = 1,2,3,\ldots . \tag{3.5.4}$$

If $t = 1$, then $t_n = 0$ for all $n = 1,2,3,\ldots$ and $[3^n] - 3[3^{n-1}] = 0$. Hence, (3.5.4) holds for all $t \in [0,1]$ and all $n = 1,2,3,\ldots$. If $t < 1$, then $t_0 = 0$ and $[t] - 3[t/3] = 0$. If $t = 1$, then $t_0 = 1$ and $[1] - 3[1/3] = 1$ and we see that (3.5.4) is also valid for $t \in [0,1]$ and $n = 0$. Hence (3.5.4) holds for all $t \in [0,1]$ and all $n = 0,1,2,3\ldots$. With the operator k defined as in Section 3.1, $kt_j = 2 - t_j$, and with k^ν denoting the ν-th iterate of k, we have

$$k^\nu t_j = 1 + (-1)^\nu (t_j - 1). \tag{3.5.5}$$

If we denote the coordinates of the image point under the mapping (3.5.2) by $(0_{\dot{3}}b_1b_3b_5\ldots, c_{0\dot{3}}c_2c_4c_6\ldots)$, we obtain from (3.5.2) and (3.5.5) that

$$\begin{aligned} b_{2n-1} &= k^{t_0+t_2+\cdots+t_{2n-2}}(t_{2n-1}-1) \\ c_0 &= t_0 \\ c_{2n} &= k^{t_1+t_3+\cdots+t_{2n-1}}t_{2n} = 1 + (-1)^{t_1+t_3+\cdots+t_{2n-1}}(t_{2n}-1) \end{aligned}$$

for $n = 1,2,3,\ldots$. From (3.5.4),

$$t_{2n-1} - 1 = [3^{2n-1}t] - 3[3^{2n-2}t] - 1$$

and

$$t_m = [3^m t] + 3[3^{m-1}t] (\text{mod } 2).$$

Hence,

$$\begin{aligned} b_{2n-1} &= 1 + ([3^{2n-1}t] - 3[3^{2n-2}t] - 1) \\ &\quad \times (-1)^{[t]+[3^2t]+3[3t]+[3^4t]+3[3^3t]+\cdots+[3^{2n-2}t]+3[3^{2n-3}t]} \\ &= 1 + ([3^{2n-1}t] - 3[3^{2n-2}t] - 1)(-1)^{[t]+[3t]+[3^2t]+\cdots+[3^{2n-2}t]}. \end{aligned}$$

Similarly,

$$c_{2n} = 1 + ([3^{2n}t] - 3[3^{2n-1}t] - 1)(-1)^{[t]+[3t]+[3^2t]+\cdots+[3^{2n-1}t]}$$

for $n = 1,2,3,\ldots$.

Finally,

$$c_0 = [t].$$

With $b_{2n-1} = f_{2n-1}(t)$ and $c_{2n} = f_{2n}(t)$, we see that our version of Cesàro's representation (3.5.3), (3.5.3a) is verified. That it does yield the correct result for $t = 1$ may be seen as follows: From (3.5.3a)

$$\begin{aligned} f_m(1) &= 1 + ([3^m] - 3[3^{m-1}] - 1)(-1)^{1+3+9+\cdots+3^{m-1}} \\ &= 1 - (-1)^{(3^m-1)/2} = \begin{cases} 0 & \text{if m is even} \\ 2 & \text{if m is odd} \end{cases}. \end{aligned}$$

Since $t = 1$, we have $[t] = 1$ and, hence,

$$\varphi_p(1) = \sum_{j=1}^{\infty} 2/3^j = 0.\overline{2} = 1, \psi_p(1) = 1 + \sum_{j=1}^{\infty} 0/3^j = 1.$$

3.6. Approximating Polygons for the Peano Curve

As in Definition 2.6, we call the polygonal line that joins the nodes i.e., the image points $f_p(0), f_p(1/3^{2n}), f_p(2/3^{2n}), \ldots, f_p((3^{2n-1})/3^{2n}), f_p(1)$ the *nth approximating polygon for the Peano curve.* The formal definition is obtained from (2.6.1) by replacing 2^{2n} by 3^{2n} and f_h by f_p wherever they occur. The second approximating polygon is depicted in Fig. 3.6.1. This drawing is not particularly helpful because the polygon keeps bumping into itself which obscures its progression, nor is it inspiring or attractive. To avoid this, we have rounded off the corners and obtained representations of the first and second approximating polygons that are depicted in Fig. 3.6.2.

As in Section 2.6, we see that these polygons form a sequence (of continuous curves) that converges uniformly to the Peano curve, providing an alternate proof of the continuity of the Peano curve. (See also

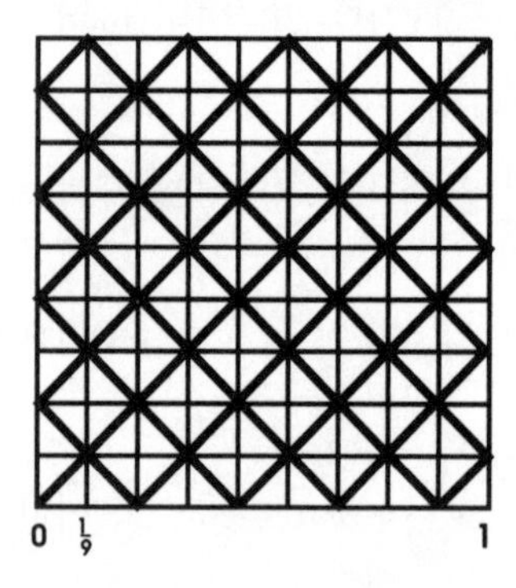

Fig. 3.6.1. Second Approximation Polygon for the Peano Curve

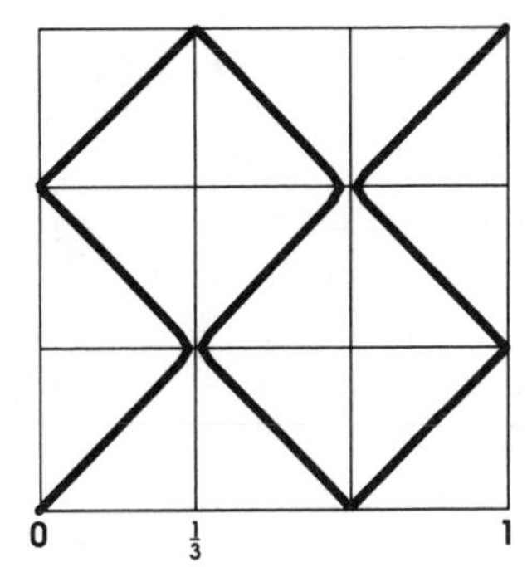

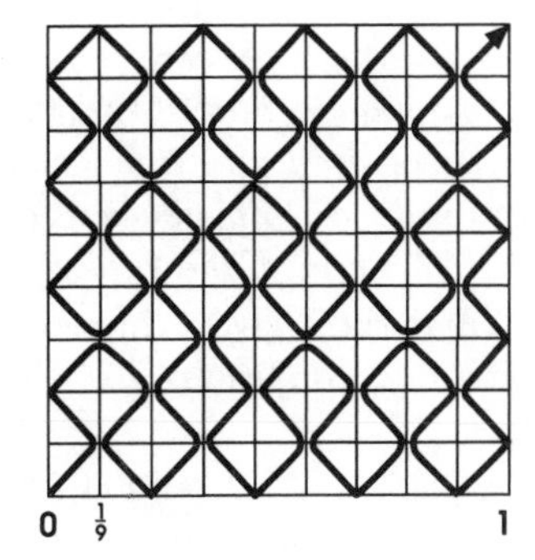

Fig. 3.6.2. First and Second Approximating Polygon for the Peano Curve with the Corners Rounded Off

Appendix A.2.4.) The approximating polygons may be generated by applying the transformations (3.4.1) to the diagonal from $(0,0)$ to $(1,1)$ as Leitmotiv.

3.7. Wunderlich's Versions of the Peano Curve

Walter Wunderlich (b.1910) was born in Vienna, Austria. He attended the University of Vienna, where he studied mathematics under Furtwängler, Mayrhofer, and Wirtinger, and the University of Technology in Vienna, where he studied geometry under Eckhart, Krames, and Kruppa. He received his doctor's degree from the latter institution in 1934 and joined the faculty in 1946. His many contributions, disseminated in 206 publications, fall mainly into two categories: classical differential geometry in euclidean and noneuclidean spaces, and applied geometry, which encompasses descriptive geometry, kinematics, and geodesy. He is the author of a widely used two-volume work on descriptive geometry and a book on kinematics. In 1970, he served as visiting professor for kinematics at Washington State University. He has served the mathematical community well, holding down the editorship of the International Mathematical News from 1947 to 1977. From 1957 to 1959, he was dean of the College of Natural

Sciences at Vienna's University of Technology, and from 1964 to 1965 he was rector magnificus of that institution. Since 1954 he has vigorously promoted the use of similarity transformations (Iterated Function Systems) for the generation of certain nowhere differentiable curves and, in particular, of space-filling curves of the Hilbert type. (For more details, see Stachel [1].)

W. Wunderlich gives three other enumerations of the subsquares that satisfy our conditions for the generation of a space-filling curve (Wunderlich [3]).

In Figs. 3.7.1, 3.7.2, and 3.7.3, we have illustrated the second and third steps in the generation of these curves. Wunderlich calls the space-filling curves obtained by the generating processes in Figures 3.7.1 and 3.7.2 *Peano curves of the switch-back type (Serpentinentyp)* and the one obtained from the generating process in Fig. 3.7.3 a *Peano curve of the meander type.* Note that the original Peano curve of Section 3.1 is also of the switch-back type. Since, in the construction of the polygons of the

Fig. 3.7.1. Generation of a Peano Curve of the Switch-back Type

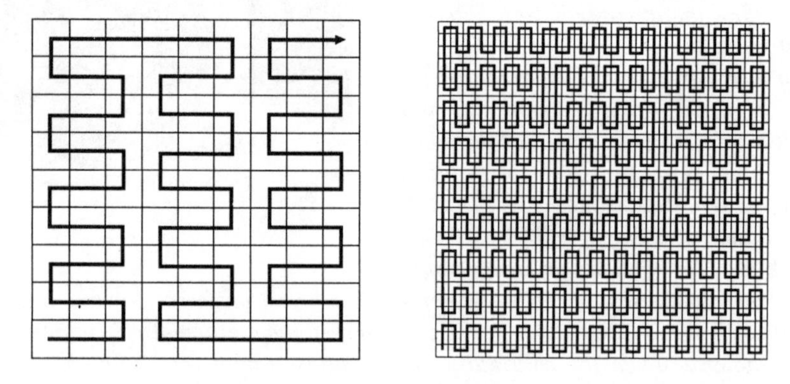

Fig. 3.7.2. Generation of a Peano Curve of the Switch-back Type

Fig. 3.7.3. Generation of a Peano Curve of the Meander Type

switch-back type, the repetition of the preceding one can be defined at every step in two different ways, one can characterize the construction process by a nine-digit binary sequence. There are $2^9 = 512$ such sequences. Among them, $2^5 = 32$ are symmetric with respect to the element in the center of the sequence because there are two ways of choosing the element in the center and 2^4 ways of choosing the four elements preceding it. Among the remaining $512 - 32 = 480$ asymmetric sequences, only half lead to different constructions, because the reversal of such a sequence effects a rotation of 180° of the square about its center. Hence, there are, altogether, $240+32 = 272$ distinct Peano curves of the switch-back type. There are only two Peano curves of the meander type because there are two different ways to start out but, from then on, there is only one mode of repetition possible. The two are, however, symmetric to each other with respect to the main diagonal of the square. Note that the curve of the meander type emanates from $(0,0)$ and terminates at $(1,0)$ and not at $(1,1)$. (Its symmetric image terminates at $(0,1)$.)

3.8. A Three-Dimensional Peano Curve

Peano defines the components $\varphi_p, \psi_p, \chi_p$ of a mapping $f_p : \mathcal{I} \to \mathcal{W}$ as follows:

$$\begin{aligned}
\varphi_p(0_{\dot{3}}t_1t_2t_3\ldots) &= 0_{\dot{3}}\beta_1\beta_2\beta_3\ldots \\
\psi_p(0_{\dot{3}}t_1t_2t_3\ldots) &= 0_{\dot{3}}\gamma_1\gamma_2\gamma_3\ldots \\
\chi_p(0_{\dot{3}}t_1t_2t_3\ldots) &= 0_{\dot{3}}\delta_1\delta_2\delta_3\ldots,
\end{aligned}$$

where

$$\beta_1 = t_1, \qquad \beta_2 = k^{\gamma_1+\delta_1} t_4, \qquad \beta_3 = k^{\gamma_1+\gamma_2+\delta_1+\delta_2} t_7$$
$$\gamma_1 = k^{\beta_1} t_2, \qquad \gamma_2 = k^{\beta_1+\beta_2+\delta_1} t_5, \qquad \gamma_3 = k^{\beta_1+\beta_2+\beta_3+\delta_1+\delta_2} t_8$$
$$\delta_1 = k^{\beta_1+\gamma_1} t_3, \qquad \delta_2 = k^{\beta_1+\beta_2+\gamma_1+\gamma_2} t_6, \qquad \delta_3 = k^{\beta_1+\beta_2+\beta_3+\gamma_1+\gamma_2+\gamma_3} t_9$$

and, in general,

$$\begin{aligned}\beta_n &= k^{\gamma_1+\cdots+\gamma_{n-1}+\delta_1+\cdots+\delta_{n-1}} t_{3n-2}\\ \gamma_n &= k^{\beta_1+\cdots+\beta_n+\delta_1+\cdots+\delta_{n-1}} t_{3n-1}\\ \delta_n &= k^{\beta_1+\cdots+\beta_n+\gamma_1+\cdots+\gamma_n} t_{3n}, \qquad n = 1, 2, 3, \ldots,\end{aligned}$$

where $kt_j = 2 - t_j$ (Peano [1]). To show that this mapping is surjective, we have to solve the above (defining) equations for $t_1, t_2, t_3, \ldots$. Since k is its own inverse, we obtain

$$\begin{aligned}t_{3n-2} &= k^{\gamma_1+\cdots+\gamma_{n-1}+\delta_1+\cdots+\delta_{n-1}} \beta_n\\ t_{3n-1} &= k^{\beta_1+\cdots+\beta_n+\delta_1+\cdots+\delta_{n-1}} \gamma_n\\ t_{3n} &= k^{\beta_1+\cdots+\beta_n+\gamma_1+\cdots+\gamma_1+\cdots+\gamma_n} \delta_n, \qquad n = 1, 2, 3, \ldots .\end{aligned}$$

The continuity may be established along the same lines as in Section 3.1 for the two-dimensional Peano curve and is left to the reader. We also leave it to the reader to prove that this curve is nowhere differentiable and to formulate a geometric generating procedure. Four- and higher-dimensional Peano curves will be encountered in Section 6.9.

3.9. Problems

1. Find $f_p(0_{\dot{3}}00001)$, $f_p(0_{\dot{3}}33333)$, $f_p(0_{\dot{3}}310221)$, $f_p(0_{\dot{3}}310\overline{2})$ from (3.1.1).
2. Show that the first component φ_p of f_p (as defined in (3.1.1)) yields the same value regardless of whether $t = 0_{\dot{3}}t_1t_2t_3\ldots t_{2n-1}t_{2n}$ or $t = 0_{\dot{3}}t_1t_2t_3\ldots t_{2n-1}(t_{2n}-1)\overline{2}$.
3. Show that the second component ψ_p of f_p (as defined in (3.1.1)) yields the same value regardless of whether $t = 0_{\dot{3}}t_1t_2t_3\ldots t_nt_{n+1}$ or $t = 0_{\dot{3}}t_1t_2t_3\ldots t_n(t_{n+1}-1)\overline{2}$.
4. Show that (3.1.1) maps the intervals $[\frac{j-1}{9}, \frac{j}{9}]$, $j = 3, 4, 5, 6, 7, 8, 9$ into squares 3,4,5,6,7,8,9 in Fig. 3.1.
5. Show that $f_p(0_{\dot{3}}t_1t_2) = g_p(0_{\dot{3}}t_1t_2)$ for $(t_1, t_2) = (0,1), (0,2), (1,0), (1,1), (2,0), (2,1), (2,2)$.
6. Show that $f_p(0_{\dot{3}}t_1t_2t_3\ldots t_{2n}) = g_p(0_{\dot{3}}t_1t_2t_3\ldots t_{2n})$ for $(t_1, t_2) = (0,0), (0,1), (1,0), (1,1), (1,2), (2,0), (2,2)$, given (3.4.4).
7. Use Cesàro's representation (3.5.3) to find $f_p(\frac{1}{3})$ and $f_p(\frac{7}{9})$.
8. Generate Peano curves of the switchback type other than the original Peano curve and the ones in Figs. 3.7.1 and 3.7.2.

9. Find the geometric generating process for the three-dimensional Peano curve of Section 3.7.
10. Let $f : \mathcal{I} \to \mathcal{Q}$ be defined by $f(0_{\dot{b}}\beta_1\beta_2\beta_3\ldots) = \begin{pmatrix} 0_{\dot{b}}\beta_1\beta_3\beta_5\ldots \\ 0_{\dot{b}}\beta_2\beta_4\beta_6\ldots \end{pmatrix}$. Show that f is surjective but not continuous.
11. Show that the three-dimensional Peano curve of Section 3.8 is continuous.
12. Show that the three-dimensional Peano curve of Section 3.7 is nowhere differentiable.
13. Find the catalogues of similarity transformations that will generate the Peano-type curves of Figs. 3.7.1, 3.7.2, and 3.7.3.
14. Find $\int_0^1 \varphi_p(t)dt$ and $\int_0^1 \psi_p(t)dt$.
15. Let $(\xi, \eta) \in \mathcal{Q}$. Show:
    (a) If both ξ and η have a unique ternary representation, then there is exactly one $t \in [0, 1]$ such that $\varphi_p(t) = \xi, \psi_p(t) = \eta$.
    (b) If ξ has a finite ternary representation and η has a unique (infinite) ternary representation, or vice versa, then there are two distinct values $t_1, t_2 \in [0, 1]$ such that $\varphi_p(t_i) = \xi, \psi_p(t_i) = \eta, i = 1, 2$.
    (c) If both ξ and η have finite ternary representations, then there may be four distinct values $t_1, t_2, t_3, t_4 \in [0, 1]$ such that $\varphi_p(t_i) = \xi, \psi_p(t_i) = \eta, i = 1, 2, 3, 4$.
16. Show that the set of points in $\mathcal{Q}$ which, under Peano's mapping $f_p : [0, 1] \to \mathcal{Q}$, have at least two preimages in [0,1] has the cardinality $\mathfrak{c}$ of the continuum.

Chapter 4

Sierpiński's Space-Filling Curve

4.1. Sierpiński's Original Definition

Waclaw Sierpiński (1882–1969) was born and died in Warsaw, Poland. In 1899, he entered the (Russian) University of Warsaw. He graduated in 1904 with a degree in science. He went on to teach grammar school and participated in the great school strike that was connected with the 1905 revolution. He then went on to Cracow where, at the University of Cracow, he received his doctor's degree in 1906. After teaching at a grammar school, he became docent at the University of L'vov in 1908 and professor in 1910. At that time, he also changed the focus of his studies from number theory to topology, not returning to number theory until around 1950. In his publications during the last twenty years of his life, number theory was the predominant topic. He was a most prolific mathematician, having published about 700 papers and books, about 600 of which were in topology. In 1914, he was interned by the Tsarist authorities. In Moscow, he was befriended by Egorov and Lusin, and a long period of cooperation between him and Lusin commenced. He returned to L'vov in 1918 and became professor at the reborn Polish University of Warsaw. At that time, he, together with Mazurkiewicz and Janiszewski, founded the new Polish School of Mathematics, which focused on foundations, set theory, and applications. They also founded the Fundamenta Mathematicae, of which he and Mazurkiewicz assumed the editorship after Janiszewski's death from influenza in 1920, and which he directed for many decades. Under his chief-editorship, the Acta Arithmetica was revived. During the occupation, when all institutions of higher learning were shut down, he continued

teaching clandestinely, and at great personal risk in his apartment. After the war, he was instrumental in organizing the Mathematical Institute of the Polish Academy of Science and remained in its presidium almost to the end of his life. Many honors were bestowed upon him during his lifetime. After his death, in honor of his memory, the main auditorium at the Mathematical Institute at the Polish Academy of Science, as well as a prize established by the Polish Mathematical Society, were named after him. So was one of Warsaw's streets and even one of the moon's craters. His grave stone carries (as per his request) the inscription "Badacz Nieskończoności" ("investigator of infinity"). For details, see Kuratowski ([2] and [3]).

In 1912, W. Sierpiński introduced still another space-filling curve (Sierpiński [1].) He showed that there is a bounded, continuous, and even function f of a real variable t which satisfies the functional equations

$$f(t) + f(t + 1/2) = 0 \quad \text{for all real} \quad t$$

and

$$2f(t/4) + f(t + 1/8) = 1 \quad \text{for all} \quad t \in [0, 1]$$

and that

$$\left.\begin{aligned} x &= f(t) \\ y &= f(t - 1/4) \end{aligned}\right\} \qquad 0 \le t \le 1 \tag{4.1.1}$$

passes through every point of the square $[-1, 1]^2$. In the course of a very laborious proof that this function f has all the required properties, he finds the following representation for it:

$$f(t) = \frac{\Theta(t)}{2} - \frac{\Theta(t)\Theta(\tau_1(t))}{4} + \frac{\Theta(t)\Theta(\tau_1(t))\Theta(\tau_2(t))}{8} - \cdots, \tag{4.1.1a}$$

where the 1-periodic functions Θ, τ_k are defined as follows:

$$\Theta(t) = \begin{cases} -1 & \text{if} \quad t \in [1/4, 3/4) \\ 1 & \text{if} \quad t \in [0, 1/4) \quad \text{or} \quad t \in [3/4, 1) \end{cases}$$

and

$$\tau_1(t) = \begin{cases} 1/8 + 4t & \text{if} \quad t \in [0, 1/4) \quad \text{or} \quad t \in [1/2, 3/4) \\ 1/8 - 4t & \text{if} \quad t \in [1/4, 1/2) \quad \text{or} \quad t \in [3/4, 1) \end{cases}$$

$$\tau_{k+1}(t) = \tau_k(\tau_1(t)), \quad k = 1, 2, 3, \ldots .$$

He then demonstrates that his curve is the limit of a uniformly convergent sequence of polygons, the first four of which are represented in Fig. 4.1. G. Pólya pointed out in a 1913 paper (Pólya [1], p. 1, footnote 3) that half of Sierpiński's curve lies in the one, while the other lies in the other half of the two right isosceles triangles that are obtained by slicing the square $[-1, 1]^2$ into half by its diagonal. Therefore, we may view Sierpiński's curve as a map from $\mathcal{I}$ onto a right isosceles triangle, and we will treat it as

Fig. 4.1. Approximating Polygons for the Sierpiński Curve

such, beginning with the next section, where we will generate it by applying Hilbert's geometric generating procedure. This will lead, in turn, to two analytic-arithmetic representations of this curve, both of which are simpler than Sierpiński's original one.

4.2. Geometric Generation and Knopp's Representation of the Sierpiński Curve

We will define a map $f_s : \mathcal{I} \to \mathcal{T}$ where $\mathcal{T}$ represents the isosceles right triangle with vertices at $(0,0), (2,0)$ and $(1,1)$, using Hilbert's geometric generating principle. We partition $\mathcal{I}$ into 2^n congruent subintervals and $\mathcal{T}$ into 2^n congruent subtriangles ($n = 1, 2, 3, \ldots$). We have illustrated the first eight steps in Fig. 4.2.1, where the bold polygonal lines indicate the order in which the subtriangles have to be arranged in order to satisfy our requirements that adjacent subintervals be mapped onto adjacent triangles with an edge in common and that each mapping preserve the preceding one. (Note how Sierpiński's approximating polygons in Fig. 4.1 may be obtained from Fig. 4.2.1 by combining Fig. 4.2.1 (a), (c), (e), (g),... with

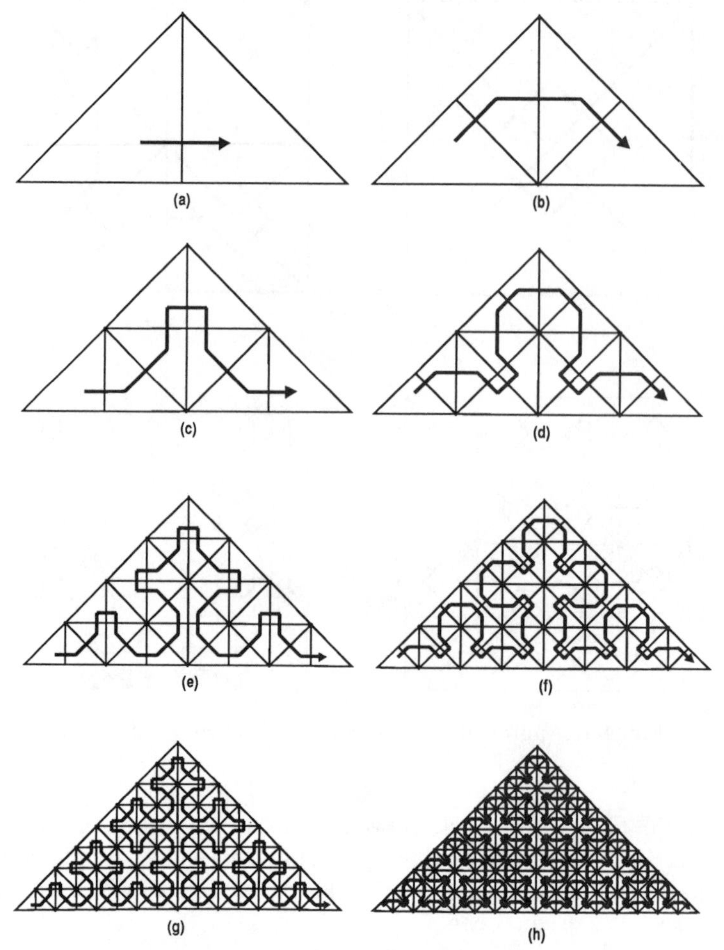

Fig. 4.2.1. Generation of the Sierpiński Curve

their mirror images with respect to their hypotenuse and a rotation through 45°.) We then define the map f_s as in Sections 2.1 or 3.3 to obtain a space-filling curve. Continuity can be established as in Section 2.1: If two points in $\mathcal{I}$ are within $1/2^n$ units from each other, then they lie, at worst, in two adjacent subintervals of the nth partition and their images lie, at worst, in two adjacent subtriangles.

Since the subtriangles shrink into points as $n \to \infty$, the curve emanates from $(0,0)$ and terminates at $(2,0)$. The requirement that the exit point from each subtriangle has to coincide with the entry point of the following one induces an orientation in each subtriangle, which we have indicated by a bold arrow in Fig. 4.2.2 for the initial triangle and the first two iterations.

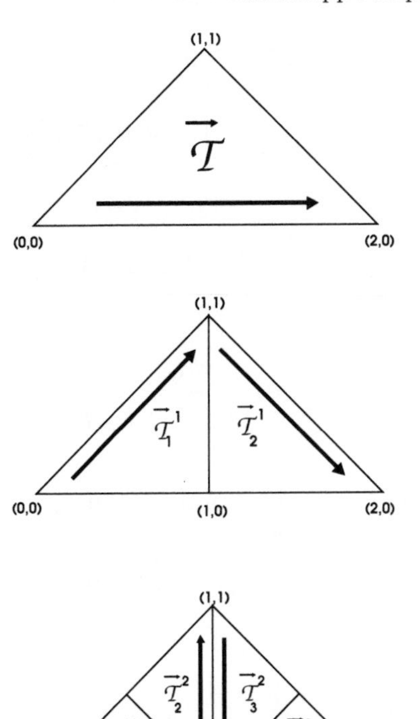

Fig. 4.2.2. Mapping $\vec{\mathcal{T}}$ onto its Congruent Parts

We see from Fig. 4.2.2 that the subtriangles of the nth partition (into 2^n subtriangles) are obtained by mapping $\vec{\mathcal{T}}$, carrying the $(n-1)$-th partition, by means of two similarity transformations $\mathfrak{K}_o$ and $\mathfrak{K}_1$ onto $\overleftarrow{\mathcal{T}}_1^1$ and onto $\vec{\mathcal{T}}_2^1$.

In order to obtain $\vec{\mathcal{T}}_1^1$ from $\vec{\mathcal{T}}$, again using complex representation, we reflect $\vec{\mathcal{T}}$ on the imaginary axis, $z' = -\overline{z}$, rotate it about the origin through $-135°$, $z'' = e^{-3\pi i/4}z'$, and shrink it uniformly towards the origin in the ratio $\sqrt{2} : 1$, $z''' = (1/\sqrt{2})z''$, to obtain

$$\mathfrak{K}_o z = -(1/\sqrt{2})e^{-3\pi i/4}\overline{z} \triangleq \frac{1}{2}K_0 z + \frac{1}{2}k_o, \tag{4.2.1}$$

where $k_0 = 0 + i0$. To obtain $\vec{\mathcal{T}}_2^1$ from $\vec{\mathcal{T}}$, we reflect $\vec{\mathcal{T}}$ on the real axis, $z' = \overline{z}$, rotate it about the origin through $-45°$, $z'' = e^{-\pi i/4}z'$, shrink it uniformly towards the origin in the ratio $\sqrt{2} : 1$, $z''' = (1/\sqrt{2})z''$, and, finally, shift it one unit to the right and one unit upwards to obtain

$$\mathfrak{K}_1 z = (1/\sqrt{2})e^{-\pi i/4}\overline{z} + \sqrt{2}e^{\pi i/4} \triangleq \frac{1}{2}K_1 z + \frac{1}{2}k_1. \tag{4.2.2}$$

We obtain, as in Section 2.3 or 3.4, that

$$f_s(0_{\dot{2}}b_1b_2b_3\ldots) = \begin{pmatrix} \mathcal{R}e \\ \mathcal{I}m \end{pmatrix} \lim_{n\to\infty} \mathfrak{K}_{b_1}\mathfrak{K}_{b_2}\mathfrak{K}_{b_3}\ldots\mathfrak{K}_{b_n}\mathcal{T}$$

and, in particular,

$$f_s(0_{\dot{2}}b_1b_2b_3\ldots b_n) = \begin{pmatrix} \mathcal{R}e \\ \mathcal{I}m \end{pmatrix} \mathfrak{K}_{b_1}\mathfrak{K}_{b_2}\mathfrak{K}_{b_3}\ldots\mathfrak{K}_{b_n}\begin{pmatrix} 0 \\ 0 \end{pmatrix}.$$

Hence,

$$f_s(0_{\dot{2}}b_1b_2b_3\ldots b_n) = \sum_{j=1}^{n}\frac{1}{2^j}\begin{pmatrix} \mathcal{R}e \\ \mathcal{I}m \end{pmatrix} K_{b_0}K_{b_1}K_{b_2}\ldots K_{b_{j-1}}k_{b_j},$$

where $K_{b_0} = 1$, and in view of the continuity of f_s,

$$f_s(0_{\dot{2}}b_1b_2b_3\ldots) = \sum_{j=1}^{\infty}\frac{1}{2^j}\begin{pmatrix} \mathcal{R}e \\ \mathcal{I}m \end{pmatrix} K_{b_0}K_{b_1}K_{b_2}\ldots K_{b_{j-1}}k_{b_j}, \tag{4.2.3}$$

where

$$K_{b_j}z = \sqrt{2}e^{(-1)^{b_j}i\pi/4}\overline{z}, k_o = 0, k_1 = 2\sqrt{2}\,e^{i\pi/4}.$$

Since $k_o = 0$, only those terms in the sum of (4.2.3) are present for which $b_j = 1$. In forming the jth term in the sum of (4.2.3), we have to take the conjugate complex of k_{b_j} $(j-1)$ times, the conjugate complex of $K_{b_{j-1}}k_{b_j}$ $(j-2)$ times, ..., and the conjugate complex of $K_{b_2}K_{b_3}\ldots K_{b_{j-1}}k_j$ one time, to arrive at

$$\frac{1}{2^j}K_{b_0}K_{b_1}K_{b_2}\ldots K_{b_{j-1}}k_{b_j} = \frac{(\sqrt{2})^j}{2^{j-1}}\exp\Big\{[(-1)^{b_1}-(-1)^{b_2} + (-1)^{b_3} - \cdots + (-1)^{j-2}(-1)^{b_{j-1}} + (-1)^{j-1}]\frac{i\pi}{4}\Big\}.$$

Hence,

$$\begin{aligned} &\varphi_s(0_{\dot{2}}b_1b_2b_3\ldots) + i\psi_s(0.b_1b_2b_3\ldots) \\ &= \sqrt{2}b_1e^{i\pi/4} + \sum_{j=2}^{\infty}\frac{b_j}{(\sqrt{2})^{j-1}} \\ &\times e^{[(-1)^{b_1}-(-1)^{b_2}+(-1)^{b_3}-\cdots+(-1)^{j-2}(-1)^{b_j-1}+(-1)^{j-1}]\frac{i\pi}{4}}. \end{aligned} \tag{4.2.4}$$

This formula was obtained by K. Knopp in 1917 as a special case of a more general formula in a paper on a unified generation and representation of the curves by von Koch, Osgood, and Peano (where he uses the term Peano curve in the generic sense of space-filling curve and does not refer to Peano's specific example—Knopp [1].) The von Koch curve is just a nowhere differentiable curve (Koch [1]—see also Sections 8.3 and 9.1), the

Osgood curve is a nowhere differentiable Jordan curve of positive Lebesgue measure (more will be said about such curves in Chapter 8), and Knopp's Peano curve is the Sierpiński curve of this section, which is also known as the Knopp curve. We will henceforth refer to it at the Sierpiński-Knopp curve. Our formula (4.2.4) differs from Knopp's formula (Knopp [1], middle of p. 113), in some small details: First, Knopp uses a triangle of base 1 rather than 2 as we have done. This accounts for the extra factor 2 in (4.2.4). Secondly, the last term of the exponent of e in Knopp's formula—once the indicated substitutions have been made—reads $-(-1)^{j-1}(-1)^{b_j}$ rather than $+(-1)^{j-1}$ as in (4.2.4) (see also Sagan [6]). This is the same, however, since only terms with $b_j = 1$ are present in (4.2.4).

Equation (4.2.4) is simpler and more easily accessible than (4.1.1) in conjunction with (4.1.1a). In the next section, we will obtain an even simpler representation by taking the iteration two steps at a time and proceeding as in Section 2.4.

Konrad Knopp (1881–1957) was born in Berlin and died in Annecy, France. In 1901, he enrolled at the University of Lausanne but, after one semester, returned to Berlin, where he completed his studies. He received his doctor's degree from the University of Berlin in 1907, having written a thesis under the direction of E. Landau, wherein he demonstrated that a divergent sequence that has a Hölder "limit" also has a Cesàro "limit." (Later, W. Schnee showed that the converse is also true, establishing the equivalence of the two concepts.) In 1908, he went to Japan, via the United States of America, where he taught at the Business School in Nagasaki. From there, he traveled to China and India. By the time he returned to Germany in 1910, he had, at the age of 28, already accomplished a complete circumnavigation of the globe. He went on to China, where he taught briefly at the German-Chinese University at Tsingtau. In 1911, he returned to the University of Berlin, where he qualified as Privatdozent. He also taught part-time at the Military Academy and the War College. In 1914, he was called up as a reserve officer to serve in World War I. The same year, wounded and decorated, he returned to teaching. In 1915, he went to teach at the University of Königsberg and, in 1926, he went to the University of Tübingen where he remained until his retirement in 1950.

Beginning with his doctoral thesis, Knopp devoted much of his time to studying limit processes. He introduced the concept of the "kernel of a sequence of complex numbers" as the intersection of those closed convex sets that contain almost all elements of the sequence. The "size" of the kernel establishes a measure for the strength of divergence. (If the kernel shrinks into a point, the sequence converges.) He also subjected Euler's summation method to an exhaustive study. This method is now referred to as the Euler-Knopp process. His textbook on the theory of functions of a complex variable (first published as two volumes in 1918 and later expanded into five volumes and translated into English) is well known to many generations of mathematics students here and abroad. His "Theory and Applications of Infinite Series" is a classic. With L. Lichtenstein and I. Schur, he founded the "Mathematische Zeitschrift" in 1918 and remained on its editorial board until 1957. (He served as editor-in-chief from 1934 to 1952.) From 1934 to 35 he was the editor of the "Jahresbericht der Deutschen Mathematikervereinigung." From 1945 to 46 he was dean of the College of Mathematical and Physical Sciences at the University in Tübingen. After 1945, he was President of the German Mathematical Society. In an obituary, published in the Jahresbericht d. DMV, E. Kamke and K. Zeller summed up his life as follows: "Es war ein reich mit Werten und Leistung erfülltes Leben." ("His life was rich in values and accomplishments.") (For more details, see Freudenthal [2] and Kamke and Zeller [1].)

4.3. Representation of the Sierpiński-Knopp Curve in Terms of Quaternaries

In Sagan [6], we have obtained a simpler representation of the Sierpiński-Knopp curve by skipping every other step in its generation and going from $\vec{\mathcal{T}}$ immediately to the configuration in Fig. 4.2.1(b) and from there to Fig. 4.2.1(d), etc... . In other words: we partition $\mathcal{I}$ into $4, 16, 64, \ldots$ subintervals and dissect $\vec{\mathcal{T}}$ accordingly. If we denote the similarity transformations from $\vec{\mathcal{T}}$ onto $\vec{\mathcal{T}}_j^2$ (see Fig. 4.2.2) by $\mathfrak{S}_0, \mathfrak{S}_1, \mathfrak{S}_2$, and $\mathfrak{S}_3$, we obtain from (4.2.1) and (4.2.2) that

$$\begin{aligned}
\mathfrak{S}_0 z &= \mathfrak{K}_0(\mathfrak{K}_0 z) = \mathfrak{K}_0[-\tfrac{1}{\sqrt{2}}\exp(-3\pi i/4)\bar{z}] = z/2 \\
\mathfrak{S}_1 z &= \mathfrak{K}_0(\mathfrak{K}_1 z) = \mathfrak{K}_0[\tfrac{1}{\sqrt{2}}\exp(-\pi i/4)z + \sqrt{2}\exp(\pi i/4)] = zi/2 + 1 \\
\mathfrak{S}_2 z &= \mathfrak{K}_1(\mathfrak{K}_0 z) = \mathfrak{K}_1[-\tfrac{1}{\sqrt{2}}\exp(-3\pi i/4)\bar{z}] = -zi/2 + 1 + i \\
\mathfrak{S}_3 z &= \mathfrak{K}_1(\mathfrak{K}_1 z) = \mathfrak{K}_1[\tfrac{1}{\sqrt{2}}\exp(-\pi i/4)z + \sqrt{2}\exp(\pi i/4)] + \sqrt{2}\exp(\pi i/4) \\
&= z/2 + 1.
\end{aligned}$$

Decomposing into real and imaginary parts, but retaining the notation $\mathfrak{S}_j$, we obtain the following catalogue of similarity transformations:

$$
\begin{aligned}
\mathfrak{S}_0\begin{pmatrix}\xi\\ \eta\end{pmatrix} &= \frac{1}{2}\begin{pmatrix}1 & 0\\ 0 & 1\end{pmatrix}\begin{pmatrix}\xi\\ \eta\end{pmatrix} + \frac{1}{2}\begin{pmatrix}0\\ 0\end{pmatrix} \triangleq \frac{1}{2}S_0x + \frac{1}{2}s_0\\
\mathfrak{S}_1\begin{pmatrix}\xi\\ \eta\end{pmatrix} &= \frac{1}{2}\begin{pmatrix}0 & -1\\ 1 & 0\end{pmatrix}\begin{pmatrix}\xi\\ \eta\end{pmatrix} + \frac{1}{2}\begin{pmatrix}2\\ 0\end{pmatrix} \triangleq \frac{1}{2}S_1x + \frac{1}{2}s_1\\
\mathfrak{S}_2\begin{pmatrix}\xi\\ \eta\end{pmatrix} &= \frac{1}{2}\begin{pmatrix}0 & 1\\ -1 & 0\end{pmatrix}\begin{pmatrix}\xi\\ \eta\end{pmatrix} + \frac{1}{2}\begin{pmatrix}2\\ 2\end{pmatrix} \triangleq \frac{1}{2}S_2x + \frac{1}{2}s_2\\
\mathfrak{S}_3\begin{pmatrix}\xi\\ \eta\end{pmatrix} &= \frac{1}{2}\begin{pmatrix}1 & 0\\ 0 & 1\end{pmatrix}\begin{pmatrix}\xi\\ \eta\end{pmatrix} + \frac{1}{2}\begin{pmatrix}2\\ 0\end{pmatrix} \triangleq \frac{1}{2}S_3x + \frac{1}{2}s_3.
\end{aligned}
\tag{4.3.1}
$$

(We will see in Section 9.3 that $\mathcal{T}$ is the unique "fixed point" of the mapping $\mathfrak{S}_* : K(\mathbb{E}^2) \to K(\mathbb{E}^2)$, where $K(\mathbb{E}^2)$ is the space of all non-empty compact subsets of $\mathbb{E}^2$, which is defined by $\mathfrak{S}_*(\mathcal{A}) = \bigcup_{i=0}^{3} \mathfrak{S}_{i^*}(\mathcal{A}), \mathcal{A} \in K(\mathbb{E}^2)$.)

We now proceed as in Section 2.4 to obtain with $S_{q_0} = I$,

$$f_s(0_4q_1q_2q_3\ldots) = \sum_{j=1}^{\infty} \frac{1}{2^j} S_{q_0}S_{q_1}S_{q_2}\ldots S_{q_{j-1}}s_{q_j}. \tag{4.3.2}$$

Noting that $S_0 = S_3 = I$, $S_2 = -S_1$, $S_1^2 = -I$, $S_1^3 = -S_1$, and $S_1^4 = I$, we may rewrite (4.3.2) as follows:

$$f_s(0_4q_1q_2q_3\ldots) = \sum_{j=1}^{\infty} \frac{1}{2^j}(-1)^{\eta_j} S_1^{\delta_j} s_{q_j}, \tag{4.3.3}$$

where

$$\eta_j = \text{number of 2's preceding } q_j \pmod 2 \tag{4.3.3a}$$

and

$$\delta_j = \text{number of 1's and 2's preceding } q_j \pmod 4. \tag{4.3.3b}$$

(The highest power of S_1 that can appear in (4.3.3) is the third power!) We leave it to the reader to prove that (4.3.3) yields the same result regardless of whether t is represented by an infinite quaternary with infinitely many trailing 3's or by the corresponding finite quaternary.

By checking all the possibilities $\delta_j = 0, 1, 2, 3$ for each $j = 1, 2, 3$ and noting that $q_j = 0$ leads to $\begin{pmatrix}0\\ 0\end{pmatrix}$, it is easy to verify that (4.3.3) may be written as

$$
\begin{aligned}
f_s(0_4q_1q_2q_3\ldots) &= \sum_{j=1}^{\infty} \frac{(-1)^{\eta_j}}{2^j}\operatorname{sgn}(q_j)\\
&\times \begin{pmatrix}(1-\delta_j)(1+(-1)^{\delta_j}) + \frac{1}{2}(\delta_j - 2)(1-(-1)^{\delta_j})(1+(-1)^{q_j})\\ (2-\delta_j)(1-(-1)^{\delta_j}) + \frac{1}{2}(1-\delta_j)(1+(-1)^{\delta_j})(1+(-1)^{q_j})\end{pmatrix},
\end{aligned}
\tag{4.3.4}
$$

where η_j, δ_j are as in (4.3.3a) and (4.3.3b).

Equation (4.3.4) is much more easily accessible to numerical evaluation than either Sierpiński's formula (4.1.1a) or Knopp's formula (4.2.4). We have listed a computer program for the evaluation of (4.4.3) in Appendix A.1.3.

4.4. Nowhere Differentiability of the Sierpiński-Knopp Curve

The nowhere differentiability of the coordinate functions of the Sierpiński-Knopp curve may be established, as we did for the nowhere differentiability of the coordinate functions of the Hilbert curve in Section 2.2, with the help of some geometric insight, and the reader is encouraged to do just that. We will, by contrast, present an analytic proof (Sagan [7]) that is based on the representation (4.3.4) and is similar to the analytic proof of the nowhere differentiability of the coordinate functions of the Hilbert curve, which we presented in Section 2.5.

First, let us introduce some abbreviating notation for the first component of the bulky vector under the sum of (4.3.4):

$$d(\delta_j, q_j) = (1 - \delta_j)(1 + (-1)^{\delta_j}) + \frac{1}{2}(\delta_j - 2)(1 - (-1)^{\delta_j})(1 + (-1)^{q_j}).$$

For $q_j = 1$ or 3, the second term of $d(\delta_j, q_j)$ vanishes and we obtain

$$d(\delta_j, \text{odd}) = (1 - \delta_j)(1 + (-1)^{\delta_j})$$

and, hence,

$$d(\text{odd}, \text{odd}) = 0, d(0, \text{odd}) = 2, d(2, \text{odd}) = -2. \tag{4.4.1}$$

We are now ready to show the following:

(4.4) Theorem. *The coordinate functions of the Sierpiński-Knopp curve—as represented in (4.3.4)—are nowhere differentiable.*

Proof. (We carry out the proof for the first component φ_s of f_s. The proof for the second component ψ_s runs along the same lines.) We have to distinguish three cases:

(1) Infinitely many 1's or 2's (or both) are present in the quaternary expansion of $t = 0_4 q_1 q_2 q_3 \ldots$.

Let q_{n_1} represent the first digit after the quaternary point that is either 1 or 2 and is preceded by an even number of 1's or 2's, let q_{n_2} represent the second digit after the quaternary point that is 1 or 2 and is preceded by an even number of 1's or 2's, etc.... Let

$$\underline{t}_{n_1} = 0_4 q_1 q_2 q_3 \ldots q_{n_1 - 1} 0, \quad \bar{t}_{n_1} = 0_4 q_1 q_2 q_3 \ldots q_{n_1 - 1} 3$$
$$\underline{t}_{n_2} = 0_4 q_1 q_2 q_3 \ldots q_{n_2 - 1} 0, \quad \bar{t}_{n_2} = 0_4 q_1 q_2 q_3 \ldots q_{n_2 - 1} 3, \text{ etc} \ldots .$$

The two sequences $\{\underline{t}_{n_j}\}$ and $\{\bar{t}_{n_j}\}$ satisfy the requirements of the condition for non-differentiability in Appendix A.2.1. We obtain

$$\bar{t}_{n_j} - \underline{t}_{n_j} = 3/4^{n_j}$$

and

$$\varphi_s(\bar{t}_{n_j}) - \varphi_s(\underline{t}_{n_j}) = \frac{(-1)^{\eta_{n_j}}}{2^{n_j}} d(\delta_{n_j}, 3).$$

By (4.4.1),

$$|(\varphi_s(\bar{t}_{n_j}) - \varphi_s(\underline{t}_{n_j}))/(\bar{t}_{n_j} - \underline{t}_{n_j})| = 2^{n_j+1}/3 \longrightarrow \infty,$$

and the nowhere differentiability follows (see Appendix A.2.1).

(2) There are only finitely many 1's and 2's in the quaternary expansion of $t = 0_{\dot{4}} q_1 q_2 q_3 \ldots q_n \ldots$ but infinitely many 3's are present. This includes the case of all finite quaternaries—other than 0—which may be converted to infinite quaternaries with infinitely many trailing 3's and, in particular, $t = 1 = 0_{\dot{4}}\bar{3}$.

Let n denote the smallest integer such that $q_{n+j} = 0$ or 3 for all $j = 1, 2, 3 \ldots$.

(a) There are an even number of 1's and 2's present among $q_1, q_2, \ldots, q_n$. Let n_1 denote the smallest integer such that $q_{n+n_1} = 3$, let n_2 denote the next following integer such that $q_{n+n_2} = 3$, etc... ., and let

$$t_{n_j} = 0_{\dot{4}} q_1 q_2 q_3 \ldots q_{n+n_{j-1}} 0 q_{n+n_j+1} \cdots .$$

Then,

$$|t - t_{n_j}| = 3/4^{n+n_j}.$$

Since 0 and 3 have the same effect on η_j and δ_j, i.e., no effect at all, we obtain

$$\varphi_s(t) - \varphi_s(t_{n_j}) = \frac{(-1)^{n+n_j}}{2^{n+n_j}} d(\delta_{n+1}, 3),$$

and, hence,

$$|(\varphi_s(t) - \varphi_s(t_{n_j}))/(t - t_{n_j})| = 2^{n+n_j+1}/3 \longrightarrow \infty \quad \text{as} \quad j \longrightarrow \infty.$$

(b) There are an odd number of 1's and 2's present among $q_1, q_2, \ldots, q_n$. Let $n_1, n_2, \ldots$ be as in (a) and let

$$t_{n_j} = 0_{\dot{4}} q_1 q_2 q_3 \ldots q_{n+n_j-2} 1 q_{n+n_j} \cdots \quad \text{for} \quad j \geq 2.$$

Now an even number of 1's or 2's precede the remaining digits (which are 0's or 3's). We have, from (4.4.1) and because all the η_j are the same for $j \geq n + n_2$,

$$\varphi_s(t_{n_j}) = \varphi_s(t) \pm (2/2^{n+n_j} + 2/2^{n+n_{j+1}} + 2/2^{n+n_{j+2}} + \cdots).$$

Since

$$|t - t_{n_j}| \leq 2/4^{n+n_{j-1}},$$

we have

$$|(\varphi_s(t) - \varphi_s(t_{n_j}))/(t - t_{n_j})| \geq 2^{n+n_j-2} \longrightarrow \infty \quad \text{as} \quad j \longrightarrow \infty.$$

(3) $t = 0$. Let $t_n = 3/4^n = 0_4000\ldots03$. Then

$$|0 - t_n| = 3/4^n.$$

By (4.3.3.a) and (4.3.3b), $\eta_j = \delta_j = 0$ for all j and we have

$$\varphi_s(0) - \varphi_s(t_n) = \frac{1}{2^n} d(0,3),$$

and, hence,

$$|(\varphi_s(0) - \varphi_s(t_n))/(0 - t_n)| = 2^{n+1}/3 \longrightarrow \infty \quad \text{as} \quad n \longrightarrow \infty.$$

These three cases exhaust all possibilities and our proposition is proved.

4.5. Approximating Polygons for the Sierpiński-Knopp Curve

Now that we are in a position to find the nodes of the Sierpiński-Knopp curve (by means of formula (4.3.4)), we may define the approximating polygons to that curve as in (2.7.1), with f_h to be replaced by f_s and 2^{2n} to be replaced by 2^n, wherever they occur. Specifically, the zeroth approximating polygon is the line segment joining $f_s(0)$ to $f_s(1)$ (which is the hypotenuse of the target triangle); the first approximating polygon is the broken line joining $f_s(0)$ to $f_s(1/2)$ to $f_s(1)$; the second approximating polygon joins $f_s(0)$ to $f_s(1/4)$ to $f_s(1/2)$ to $f_s(3/4)$ to $f_s(1)$, etc.... In Fig. 4.5.1 we depicted the sixth and seventh approximating polygons. As in previous cases, these figures are not very revealing, because the approximating polygons bump into themselves and double back every once in a while. Besides, they are not particularly attractive. (We may, incidentally, obtain these approximating polygons by subjecting the (directed) line segment from (0,0) to (2,0)

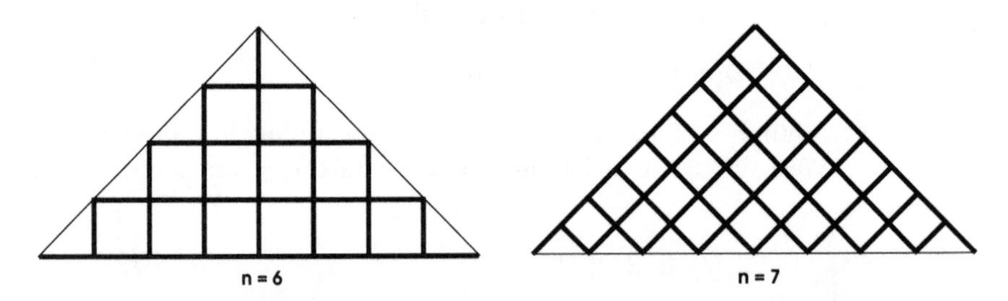

Fig. 4.5.1. Sixth and Seventh Approximating Polygon of the Sierpiński-Knopp Curve

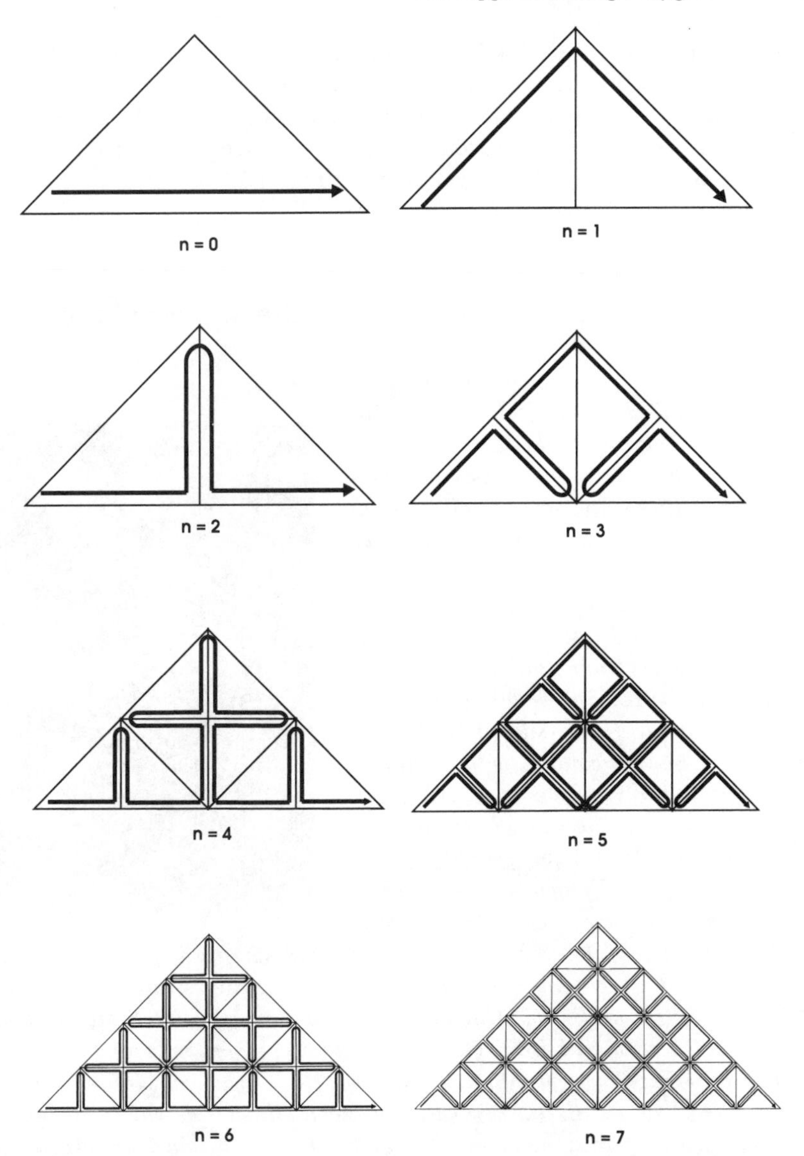

Fig. 4.5.2. Approximations to the Approximating Polygons for $n = 0, 1, 2, \ldots, 7$

repeatedly to the transformations (4.2.1) and (4.2.2). The nth approximating polygon is obtained by carrying out these transformations n times in a row.)

To obtain a better idea of the progression of the approximating polygons, we replaced them by approximations that are obtained as follows: We replace the line segment from (0,0) to (2,0) by one that is slightly above it and cut off by the remaining two sides of the target triangle $\mathcal{T}$. We then apply the transformations (4.2.1) and (4.2.2) successively. After every other step, we have to close the remaining gap between the two separate portions. We do this by means of an arc that does not quite touch the sides of the triangle to prevent the path from bumping into itself two iterations later and thereby obscuring the progression. In Fig. 4.5.2 we have represented the results of this process for $n = 0, 1, 2, \ldots, 7$.

4.6. Pólya's Generalization of the Sierpiński Curve

Georg Pólya (1887–1985) was born in Budapest and died in Palo Alto, California. He attended the University of Budapest, where he studied law, then languages and literature, and, finally, mathematics and physics under such luminaries as L. Eötvös and L. Fejér. He spent the years 1910 and 1911 in Vienna, and returned to Budapest to receive his doctor's degree from the University of Budapest in 1912. Subsequently, he spent two years at the University of Göttingen, where he came in contact with F. Klein, D. Hilbert, H. Weyl, and R. Courant. In Paris in 1914, he met É. Picard and J. Hadamard. That same year, he joined the faculty of the Eidgenössische Technische Hochschule in Zürich at the behest of A. Hurwitz. There, he was appointed to a professorship in 1928 and stayed until 1940, at which time he emigrated to the United States of America. After two years at Brown University and a short stay at Smith College, he responded to a call from G. Szegö, who was then head of the Mathematics Department at Stanford University. He taught at Stanford until his "retirement" in 1953. His contributions to mathematics education during his long years of retirement are legendary. (When he died he was almost 98 years old.) During his long professional career, he collaborated with G. Szegö (with whom he published the classic "Aufgaben und Lehrsätze aus der Analysis"—Springer-Verlag, 1925), G. Julia, I. Schur, M. Fekete, A. Hurwitz, G.H. Hardy, J.E. Littlewood, N. Wiener, I.J. Schoenberg, and many others. Among the many areas to which he contributed are the theory of functions of a complex variable, probability theory, mathematical methods, and combinatorics. (For more details, see Alexanderson [1].)

In 1913, one year after the appearance of Sierpiński's paper [1], G. Pólya published a generalization of Sierpiński's construction (Pólya [1]). While Hilbert [1] hinted that space-filling curves could be constructed so that every point in the target set has, at most, three preimages in $\mathcal{I}$ but did not elaborate, Pólya was the first to demonstrate how this can be accomplished—and therein lies the significance of his contribution.

Pólya maps $\mathcal{I}$ onto a right triangle, which need not be isosceles, by first partitioning $\mathcal{I}$ into two congruent subintervals and splitting the triangle $\mathcal{T}$ into two similar but not congruent subtriangles, as indicated in Fig. 4.6.1. He then proceeds as indicated in Fig. 4.6.2, where the bold polygonal lines indicate the order in which the subtriangles have to be taken.

Our conditions for the definition of a space-filling curve as formulated in Section 2.1 are obviously met, and we proceed as before. We note that as the ratio of the two sides approaches one, the better the polygonal lines in Fig. 4.6.2 resemble the ones in Fig. 4.2.1.

In the early 1970s, Peter D. Lax asked his students in Real Variables at New York University to prove that Pólya's curve is nowhere differentiable. Failing to elicit any response, he tried it himself and found to his surprise that Pólya's curve is nowhere differentiable if the smaller angle in Fig. 4.6.1 is between 30° and 45°, that it is nowhere differentiable on a set of measure 1 and has derivative zero on a non-denumerable set if that angle is between 15° and 30°, and that is derivative is zero on a set of measure 1 if the angle is less than 15° (Lax [1]). R.T. Bumby proved subsequently that it is nowhere differentiable if the angle is 30°, and is not differentiable on a set of measure 1 if the angle is 15° (Bumby [1]). (When the angle is 45°, Pólya's curve becomes the Sierpiński-Knopp curve and Theorem 4.4 applies.)

Pólya has shown the following:

(4.6) Theorem. *(Pólya): If one chooses the ratio of the shorter side to the hypotenuse (which is, in any event, less than $1/\sqrt{2}$) to be a transcendental number, then every point in $\mathcal{T}$ has, at most, three preimages in $\mathcal{I}$.*

Proof. Observe that every point in $\mathcal{T}$ belongs to at least one subtriangle and, at most, eight subtriangles: If P does not lie on any of the line segments that partition the triangle, or if it is one of the two vertices at the two ends of the hypotenuse, then it belongs to exactly one subtriangle and has exactly one preimage. With every iteration, the right angles of the existing partition are split into two angles: an angle of less than 45° (*sharp angle*), and the complementary angle (*blunt angle*). Hence, there cannot be more than eight

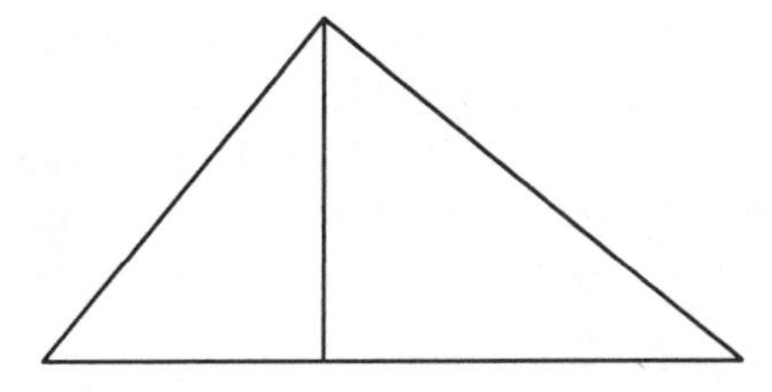

Fig. 4.6.1. Initial Step in the Generation of Pólya's Space-Filling Curve

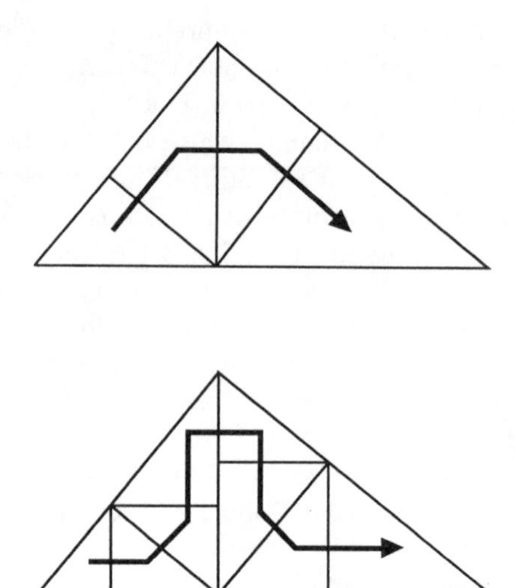

Fig. 4.6.2. Generation of Pólya's Space-Filling Curve

line segments emanating (or terminating) at a nodal point, meaning that the point cannot belong to more than eight subtriangles. No matter how many subtriangles a point belongs to, this number cannot decrease as the partition is refined, because each partition contains the preceding one. Let $\nu(P)$ denote the number of subtriangles the point P belongs to:

If $\nu(P) = 1$, then P is a simple point, i.e., has only one preimage in $\mathcal{I}$. (The point P_1 in Fig. 4.6.3 is such a point.)

If $\nu(P) = 2$ or 4, then P is either the vertex at the right angle and has only one preimage, because P (P_2 in Fig. 4.6.3) is the exit point from the triangle on the left and the entry point to the triangle on the right (the vertex at the blunt angle in the subtriangle is the entry point and the vertex at the sharp angle is the exit point), and, hence, has only one preimage, namely $t = \frac{1}{2}$, or P lies on the boundary of two triangles but not at a vertex. In the latter case, it has two preimages. All other points that appear to have $\nu(P) = 2$ are changed into points with $\nu(P) = 4$ at the very next partition. (See also Fig. 4.6.3.) When $\nu(P) = 4$, the exit point of the first triangle coincides with the entry point of the following one, giving rise to one preimage. Likewise, the exit point of the third triangle coincides with the entry point of the fourth triangle, giving rise to another preimage.

$\nu(P) = 3$ is changed into $\nu(P) = 5$ at the next partition.

If $\nu(P) = 5$, then P has three preimages in $\mathcal{I}$ (such as the point P_5 in Fig. 4.6.3). It belongs to five triangles, one of which sits completely apart

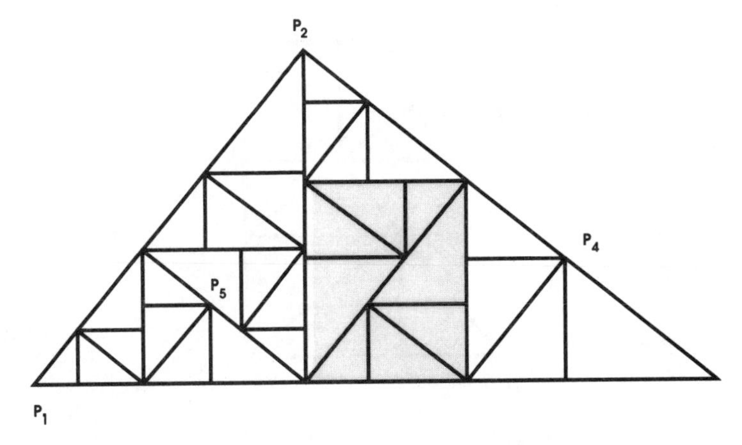

Fig. 4.6.3. Partitioning of $\mathcal{T}$ into 32 Subtriangles

from the rest, giving rise to one preimage. The remaining four are in the same position as in the preceding case, accounting for an additional two preimages.

$\nu(P) = 6$ or 7 is impossible.

If $\nu(p) = 8$, then P has four preimages. In this case, P lies on a line segment that forms the diagonal of a rectangle that is formed from two subtriangles. (See also Fig. 4.6.4, which displays the shaded rectangle from Fig. 4.6.3 after two more partitions). On either side of the diagonal lie four triangles with one of their vertices at P. This accounts for two preimages per side. We will show that this cannot happen if the ratio r of the shorter side to the hypotenuse is transcendental.

We see from Fig. 4.6.5 (where we assume the hypotenuse to have length l) and some elementary trigonometry that the distance from the beginning point 0 to P_1 is given by

$$\rho(0, P_1) = lr^2,$$

the distance from 0 to P_2 by

$$\rho(0, P_2) = lr^2 + l(1 - r^2)r^2,$$

the distance from 0 to P_3 by

$$\rho(0, P_3) = lr^2 + lr^2(1 - r^2) + lr^2(1 - r^2)^2,$$

etc.... We recognize that all these distances are expressed as the product of l with a polynomial in r with integer coefficients and no absolute term.

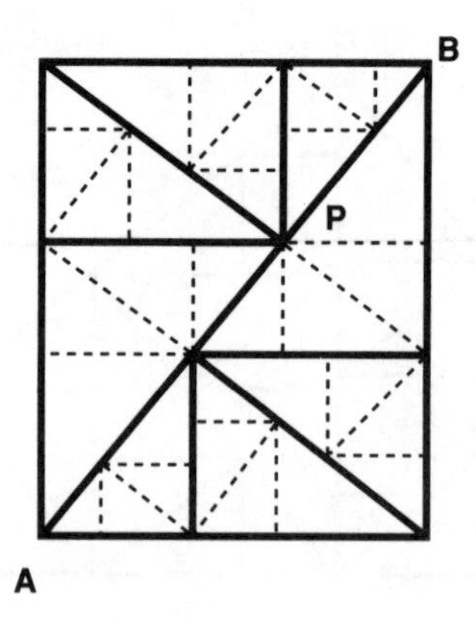

Fig. 4.6.4. A Point with $\nu(P) = 8$

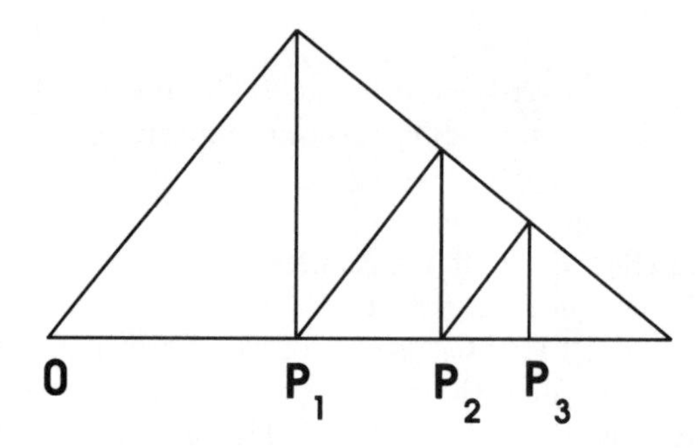

Fig. 4.6.5. Construction of the Nodes along the Hypotenuse

Therefore, if $\nu(P) = 8$, then, measuring the distance from A to P (in Fig. 4.6.4) with reference to the upper triangle, we obtain

$$\rho(A, P) = l\ pol_1(r),$$

where $pol_1(r)$ denotes a polynomial in r with integer coefficients and without absolute term and, if we measure the distance from B (with reference to the lower triangle), we obtain

$$\rho(B, P) = l\ pol_2(r),$$

where $pol_2(r)$ stands for some (other) such polynomial. Hence, we must have

$$l\ pol_1(r) + l\ pol_2(r) = l.$$

Therefore, r has to be an algebraic number. If r is transcendental, then $\nu(P) = 8$ cannot happen and our proposition is proved. □

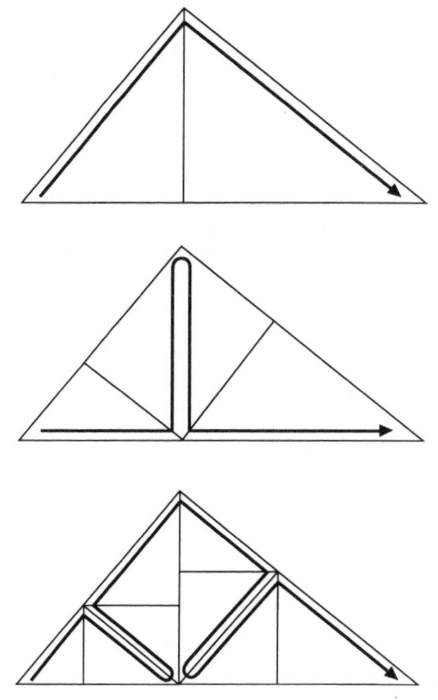

Fig. 4.6.6. Approximations to the First Three Approximating Polygons for the Pólya Curve

This is the best one can do. Hahn and Mazurkiewicz have shown that if each point of $f_*(\mathcal{I}), f : \mathcal{I} \longrightarrow \mathcal{Q}$ being continuous has, at most, two preimages in $\mathcal{I}$, then it cannot be space-filling (Hahn [1], Mazurkiewicz [4]). They went on to show that if f represents a space-filling curve, then the points in $\mathcal{Q}$ with at least three preimages in $\mathcal{I}$ are everywhere dense. Hahn also demonstrated that the set of points with at least two preimages in $\mathcal{I}$ has cardinality $\mathfrak{c}$. (Note: Hilbert's and Peano's curves have quadruple points, i.e., points with four preimages.) Unaware of Pólya's result, Hahn also demonstrated how Peano's map (3.1.1) can be modified so that each point on the curve has, at most, three preimages in $\mathcal{I}$ (Hahn [1]).

Approximations to the approximating polygons for the Pólya curve appear as we have indicated for the first three polygons in Fig. 4.6.6.

4.7. Problems

1. Present a "geometric" proof, such as the one in Section 2.2, that the Sierpiński-Knopp curve is nowhere differentiable.
2. Verify the transformation formulas (4.3.1).
3. Give a "geometric" and an analytic proof that the Sierpiński-Knopp curve is continuous.

4. Verify (4.3.4).
5. Use (4.2.4) as well as (4.3.4) to evaluate $f_s(0_{\dot{2}}1101011)$, $f_s(0_{\dot{2}}0000001)$, $f_s(0_{\dot{2}}11111111)$.
6. Show that $s_t = s_{t-1} + S_{t-1}s_3$ for $t = 1, 2, 3$.
7. Show that (4.3.3) yields the same result, regardless of whether t is represented by a finite quaternary or the equivalent infinite quaternary with infinitely many trailing 3's. (Use the result from Problem 6.)
8. Based on (4.3.4), give an analytic proof that the second component ψ_s of the Sierpiński-Knopp curve is nowhere differentiable.
9. Show that a configuration such as in Figs. 4.6.3 and 4.6.4 can only be obtained if $r = \sqrt{(3 - \sqrt{5}/2}$.
10. Show that a point P with $\nu(P) = 8$ in a configuration such as that given in Fig. 4.2.1 (d), (e),... can only appear if $r = 1/\sqrt{2}$, i.e., T is isosceles.
11. Deduce the representation (4.3.4) of the Sierpiński-Knopp curve from Sierpiński's definition in (4.1.1), (4.1.1a). (Note that in order to obtain the Sierpiński-Knopp curve from the Sierpiński curve, a rotation and scale change have to be performed and the curve has to be cut in half—see Figs. 4.1.1 and 4.2.1.)

Chapter 5

Lebesgue's Space-Filling Curve

5.1. The Cantor Set

In a paper on infinite linear point manifolds written in 1883, in which Cantor searched for a characterization of the continuum, he offers in the appendix the set of all points that can be represented by

$$\frac{2t_1}{3} + \frac{2t_2}{3^2} + \frac{2t_3}{3^3} + \frac{2t_4}{3^4} + \cdots,$$

where $t_j = 0$ or 1, as an example of a perfect set (a set that is equal to the set of all its accumulation points) that is not dense in any interval, no matter how small (Cantor [2]). This set, which had its humble beginnings as a counterexample in an appendix has since taken on a life of its own and has served ever since its inception, as an example, counterexample, and inspiration for inquiries into the most remote recesses of mathematical analysis. It is no coincidence that it appears in a fundamental rôle in the study of space-filling curves. We will see in Chapter 6 that this rôle is even more fundamental than it would appear to be from a study of the present chapter.

This set, which is now called the *Cantor Set* or the *Set of the Excluded Middle Thirds* (for reasons that will be explained later), and which we denote by Γ, may be represented by

$$\Gamma = \{0_{\dot{3}}(2t_1)(2t_2)(2t_3)\ldots \mid t_j = 0 \quad \text{or} \quad 1\}. \tag{5.1.1}$$

The function $g(0_{\dot{3}}(2t_1)(2t_2)(2t_3)\ldots) = 0_{\dot{2}}t_1t_2t_3\ldots$ defines a $1-1$ map from Γ onto the set $\mathcal{B} = \{0_{\dot{2}}b_1b_2b_3\ldots \mid b_j = 0 \quad \text{or} \quad 1\}$ of all binaries and $\mathcal{I} \subset \mathcal{B}$. On the other hand, $\Gamma \subset \mathcal{I}$ and we have

$$\mathcal{I} \subset \mathcal{B} \leftrightarrow \Gamma \subset \mathcal{I}.$$

Consequently, $\Gamma \longleftrightarrow \mathcal{I}$, i.e., the Cantor set Γ has the cardinality $\mathfrak{c}$ of the continuum.

Since $\mathcal{I} \longleftrightarrow \mathcal{Q}$ and $\mathcal{I} \longleftrightarrow \mathcal{W}$, we have that

$$\Gamma \longleftrightarrow \mathcal{Q} \quad \text{and} \quad \Gamma \longleftrightarrow \mathcal{W}.$$

It is of interest to us that there are continuous mappings from $\varGamma$ onto $\mathcal{Q}$ and onto $\mathcal{W}$, such as

$$f(0_{\dot{3}}(2t_1)(2t_2)(2t_3)\ldots) = \begin{pmatrix} 0_{\dot{2}}t_1t_3t_5\ldots \\ 0_{\dot{2}}t_2t_4t_6\ldots \end{pmatrix} \tag{5.1.2}$$

and

$$F(0_{\dot{3}}(2t_1)(2t_2)(2t_3)\ldots) = \begin{pmatrix} 0_{\dot{2}}t_1t_4t_7\ldots \\ 0_{\dot{2}}t_2t_5t_8\ldots \\ 0_{\dot{2}}t_3t_6t_9\ldots \end{pmatrix}, \tag{5.1.3}$$

where $t_j = 0$ or 1. (These mappings are, of course, *not* injective.)

The mappings $f : \varGamma \to \mathcal{Q}$ and $F : \varGamma \longrightarrow \mathcal{W}$ as defined in (5.1.2) and (5.1.3) are surjective. (We will prove this proposition for (5.1.2) only. The proof for the other case is analogous.)

Let $p \in \mathcal{Q}$. Then, the coordinates of p may be represented by the binary numbers

$$p = \begin{pmatrix} 0_{\dot{2}}a_1a_2a_3\ldots \\ 0_{\dot{2}}b_1b_2b_3\ldots \end{pmatrix}.$$

If we let $t = 0_{\dot{3}}(2a_1)(2b_1)(2a_2)(2b_2)(2a_3)(2b_3)\ldots$, then, by (5.1.2), $f(t) = p$. □

The mappings $f : \varGamma \longrightarrow \mathcal{Q}$ and $F : \varGamma \longrightarrow \mathcal{W}$ as defined in (5.1.2) and (5.1.3) are continuous. (Again, we only give a proof for (5.1.2). Suppose that $|t - t_0| < \frac{1}{3^{2n}}$. Then, t_0 and t cannot differ in the first $2n$ ternary places and we have

$$t_0 = 0_{\dot{3}}(2t_1)(2t_2)\ldots(2t_{2n})(2t_{2n+1})\ldots,$$
$$t = 0_{\dot{3}}(2t_1)(2t_2)\ldots(2t_{2n})(2\tau_{2n+1})\ldots .$$

Suppose, to the contrary, that $t_{2n} \neq \tau_{2n}$. Then, $|t_{2n} - \tau_{2n}| = 1$ and

$$t - t_0 = (2\tau_{2n} - 2t_{2n})/3^{2n} + (2\tau_{2n+1} - 2t_{2n+1})/3^{2n+1} + \cdots$$

and, hence,

$$|t - t_0| \geq (2/3^{2n}) - (2/3^{2n+1})(1 + 1/3 + 1/9 + \cdots)$$
$$= 2/3^{2n} - 1/3^{2n} = 1/3^{2n}$$

instead of $< 1/3^{2n}$. Therefore,

$$f(t) - f(t_0) = \begin{pmatrix} (\tau_{2n+1} - t_{2n+1})/2^{n+1} + (\tau_{2n+3} - t_{2n+3})/2^{n+2} + \cdots \\ (\tau_{2n+2} - t_{2n+2})/2^{n+1} + (\tau_{2n+4} - t_{2n+4})/2^{n+2} + \cdots \end{pmatrix},$$

and we have

$$\|f(t) - f(t_0)\| < \sqrt{2}/2^n,$$

whence continuity obtains. □

The mappings $f : \varGamma \longrightarrow \mathcal{Q}$ and $F : \varGamma \longrightarrow \mathcal{W}$ as defined in (5.1.2) and (5.1.3) are nowhere differentiable. (Again, the proof will only be carried out for the case (5.1.2).)

Let

$$t = 0_{\dot{3}}(2t_1)(2t_2)\ldots(2t_{2n})(2t_{2n+1})(2t_{2n+2})\ldots,$$
$$t_n = 0_{\dot{3}}(2t_1)(2t_2)\ldots(2t_{2n})(2\tau_{2n+1})(2t_{2n+2})\ldots,$$

where $\tau_{2n+1} = t_{2n+1} + 1 \pmod 2$. Then

$$|t - t_n| = 2/3^{2n+1}.$$

If φ denotes the first component of f, we have

$$\varphi(t) - \varphi(t_n) = (t_{2n+1} - \tau_{2n+1})/2^{n+1}$$

and, hence,

$$|\varphi(t) - \varphi(t_n)|/|t - t_n| = \frac{3}{4}\left(\frac{9}{2}\right)^n \longrightarrow \infty,$$

and we see that φ is nowhere differentiable on Γ. An analogous proof applies to the second component ψ of f. □

To summarize:

(5.1) Theorem. *The mappings defined in (5.1.2) and (5.2.3) are surjective, continuous, and nowhere differentiable.*

H. Lebesgue and I.J. Schoenberg have extended these mappings to the entire interval $\mathcal{I}$ to obtain space-filling curves, as we will show in Section 5.4 and Chapter 7. Hans Hahn has used them—on second thought—as a point of departure for his characterization of the most general set that can be obtained as a continuous image of a line segment. By doing this, he provided a recipe for the construction of space-filling curves, as we will see in Chapter 6.

5.2. Properties of the Cantor Set

The Cantor set, as defined in (5.1.1), may be obtained from $\mathcal{I}$ by first removing the middle third $\left(\frac{1}{3}, \frac{2}{3}\right)$ to be left with the two closed intervals $\left[0, \frac{1}{3}\right], \left[\frac{2}{3}, 1\right]$. Next, remove the middle thirds of these two intervals, namely $\left(\frac{1}{9}, \frac{2}{9}\right), \left(\frac{7}{9}, \frac{8}{9}\right)$, to be left with the four closed intervals $\left[0, \frac{1}{9}\right], \left[\frac{2}{9}, \frac{1}{3}\right], \left[\frac{2}{3}, \frac{7}{9}\right]$, and $\left[\frac{8}{9}, 1\right]$. We again remove the middle thirds of what is left and proceed with this process ad infinitum. (See Fig. 5.2.1.)
That this process leads to the Cantor set in (5.1.1) may be seen as follows: If all numbers in $\mathcal{I}$ are represented in ternary form

$$t = 0_3 t_1 t_2 t_3 \ldots,$$

where $t_j = 0$ or 1, or 2, then, by removing the middle third $\left(\frac{1}{3}, \frac{2}{3}\right)$ from $\mathcal{I}$ we remove all numbers of the form $0_3 1 t_2 t_3 t_4 \ldots$ except for $0_3 1$, which

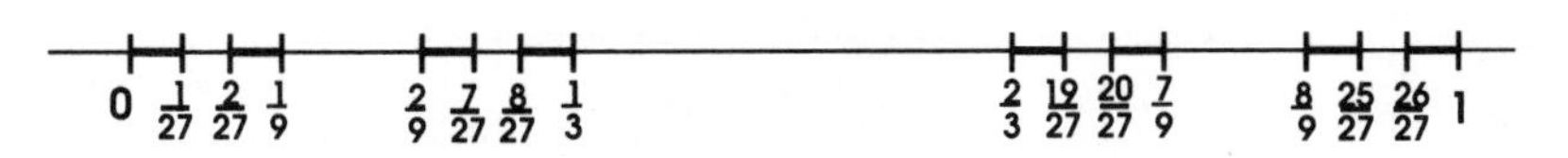

Fig. 5.2.1. Generation of the Cantor Set

we rewrite as $0_{\dot{3}}0\overline{2}$. At the next step, we remove all numbers of the form $0_{\dot{3}}01t_3t_4t_5\ldots, 0_{\dot{3}}21t_3t_4t_5\ldots$ except for $0_{\dot{3}}01$ and $0_{\dot{3}}21$, which we rewrite as $0_{\dot{3}}00\overline{2}$ and $0_{\dot{3}}20\overline{2}$. We continue in this manner and wind up with a set Γ^*, the points of which appear to have ternary representations with $0's$ and $2's$ only. In fact, it is the Cantor set as defined in (5.1.1). This follows from the following lemma:

(5.2.1) Lemma. *$t \in \Gamma^*$ if and only if at least one ternary representation of t does not contain a digit 1, i.e., $\Gamma = \Gamma^*$. (Note that every finite ternary that terminates in a 1 may be rewritten as an infinite ternary with infinitely many trailing $2's$.)*

Proof. We have seen that, for $n = 1$, an interval (a_n, b_n) is removed if $a_1 = 0_{\dot{3}}1$ and $b_1 = 0_{\dot{3}}2$. For $n = 2$, it is removed if $a_2 = 0_{\dot{3}}01$, $b_2 = 0_{\dot{3}}02$ or $a_2 = 0_{\dot{3}}21$, $b_2 = 0_{\dot{3}}22$. In general, it is removed if and only if a_n, b_n can be written in the form

$$\begin{aligned} a_n &= 0_{\dot{3}}(2a_1)(2a_2)(2a_3)\ldots(2a_{n-1})1 \\ b_n &= 0_{\dot{3}}(2a_1)(2a_2)(2a_3)\ldots(2a_{n-1})2. \end{aligned} \tag{5.2.1}$$

Suppose this is true. At the next, the $(n+1)$-th step, we have

$$a_{n+1} = b_n + 1/3^{n+1}, b_{n+1} = a_{n+1} + 1/3^{n+1}$$

(see Fig. 5.2.2) and, consequently,

$$\begin{aligned} a_{n+1} &= 0_{\dot{3}}(2a_1)(2a_2)(2a_{\dot{3}})\ldots(2a_{n-1})2 + 1/3^{n+1} \\ &= 0_{\dot{3}}(2a_1)(2a_2)92a_3)\ldots(2a_{n-1})21, \\ b_{n+1} &= a_{n+1} + 1/3^{n+1} \\ &= 0_{\dot{3}}(2a_1)(2a_2)(2a_3)\ldots(2a_{n-1})22, \end{aligned}$$

or we have

$$\begin{aligned} a_{n+1} &= a_n - \frac{2}{3^{n+1}} = 0_{\dot{3}}(2a_1)(2a_2)\ldots(2a_{n-1})1 - \frac{2}{3^{n+1}} \\ &= 0_{\dot{3}}(2a_1)(2a_2)\ldots(2a_{n-1})01 \\ b_{n+1} &= a_n - \frac{1}{3^{n+1}} = 0_{\dot{3}}(2a_1)(2a_2)\ldots(2a_{n-1})1 - \frac{1}{3^{n+1}} \\ &= 0_{\dot{3}}(2a_1)(2a_2)\ldots(2a_{n-1})02 \end{aligned}$$

and we are back to (5.2.1) with n replaced by $n+1$. Hence, the validity of (5.2.1) is established by induction.

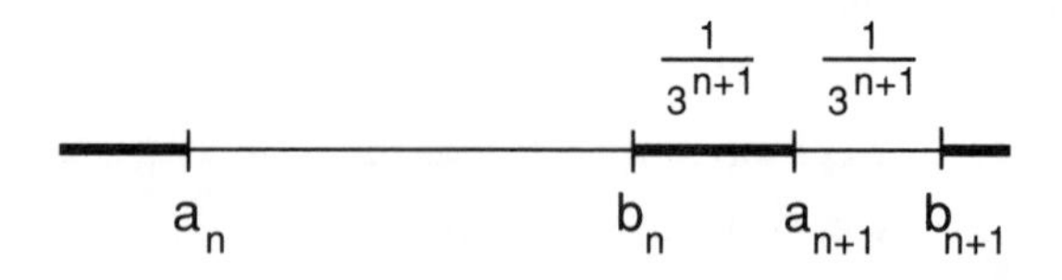

Fig. 5.2.2. Step $(n+1)$ in the Generation of the Cantor Set

If $t \in (a_n, b_n)$, then

$$t = 0_{\dot{3}}(2t_1)(2t_2)(2t_3)\dots(2t_{n-1})1\tau_{n+1}\tau_{n+2}\cdots .$$

If all $\tau_{n+j} = 0$, then $t = a_n \notin (a_n, b_n)$, or if all $\tau_{n+j} = 2$, then $t = b_n \notin (a_n, b_n)$. Therefore, at least one of the $\tau_{n+j} \neq 0$ and at least one $\tau_{n+j} \neq 2$. Hence, a point t is removed in the course of the construction of Γ^* if and only if both its ternary representations (if there are two) contain at least one digit 1, and Lemma 5.1.1 is proved. □

Let

$$\begin{aligned}
\Omega_1 &= \left(\frac{1}{3}, \frac{2}{3}\right) \\
\Omega_2 &= \left(\frac{1}{9}, \frac{2}{9}\right) \cup \left(\frac{7}{9}, \frac{8}{9}\right) \\
&\vdots \\
\Omega_n &= (a_n^{(1)}, b_n^{(1)}) \cup (a_n^{(2)}, b_n^{(2)}) \cup \ldots \cup (a_n^{(2^{n-1})}, b_n^{(2^{n-1})}), \\
&\qquad \vdots
\end{aligned} \tag{5.2.2}$$

where the $(a_n^{(k)}, b_n^{(k)})$ are the 2^{n-1} open intervals that are removed from $\mathcal{I}$ at the nth step when constructing Γ. Since all Ω_j are open, $\Gamma^c = \mathcal{I} \backslash \Gamma = \bigcup_{j=1}^{\infty} \Omega_j$ is open and Γ is *closed*, i.e., it contains all its accumulation points. In fact, Γ is a *perfect set*, i.e., it is equal to the set Γ' of its accumulation points:

(5.2.2) Lemma. $\Gamma = \Gamma'$. *Specifically, all points of the form* $0_{\dot{3}}(2t_1)(2t_2)(2t_3)\dots(2t_n)\overline{2} \in \Gamma$ *are right accumulation points, all points of the form* $0_{\dot{3}}(2t_1)(2t_2)(2t_3)\dots(2t_n) \in \Gamma$ *are left accumulation points, and all other points of* Γ *are left and right accumulation points.*

Proof. Let

$$t = 0_{\dot{3}}(2t_1)(2t_2)(2t_3)\dots(2t_n)\overline{2}$$

and

$$t^{(k)} = 0_{\dot{3}}(2t_1)(2t_2)(2t_3)\dots(2t_n)22\dots20\overline{2},$$

where the 0 (which replaces a 2) is in $(n+k)$-th position. Then, $t^{(k)} \in \Gamma$ and $\{t^{(k)}\} \longrightarrow t - 0$.

Let

$$t = 0_{\dot{3}}(2t_1)(2t_2)(2t_3)\dots(2t_n)$$

and

$$t^{(k)} = 0_{\dot{3}}(2t_1)(2t_2)(2t_3)\dots(2t_n)000\dots02,$$

where the last 2 is in $(n+k)$-th position. Then, $t^{(k)} \in \Gamma$ and $\{t^{(k)}\} \longrightarrow t+0$.

Finally, let $t = 0_{\dot{3}}(2t_1)(2t_2)(2t_3)\ldots(2t_n)\ldots$, where infinitely many $t_j = 0$ and infinitely many $t_j = 1$, and

$$t^{(n)} = 0_{\dot{3}}(2t_1)(2t_2)(2t_3)\ldots(2t_{n-1})(2\tau_n)(2t_{n+1})\ldots,$$

where $\tau_n = t_n + 1 \pmod 2$. Then, $t^{(n)} \in \Gamma$ and $\{t^{(n)}\} \longrightarrow t$. □

The Cantor set does not occupy "any space" in $\mathcal{I}$:

(5.2.3) Lemma. $\Lambda_1(\Gamma) = J_1(\Gamma) = 0$.

Proof. Since

$$\begin{aligned} \Gamma &\subset \left[0, \frac{1}{3}\right] \cup \left[\frac{2}{3}, 1\right] = \Gamma_1 \\ \Gamma &\subset \left[0, \frac{1}{9}\right] \cup \left[\frac{2}{9}, \frac{1}{3}\right] \cup \left[\frac{2}{3}, \frac{7}{9}\right] \cup \left[\frac{8}{9}, 1\right] = \Gamma_2 \\ \Gamma &\subset \left[0, \frac{1}{27}\right] \cup \left[\frac{2}{27}, \frac{1}{9}\right] \cup \ldots \cup \left[\frac{26}{27}, 1\right] = \Gamma_3 \\ &\vdots \end{aligned} \tag{5.2.3}$$

and $J_1(\Gamma_k) = \left(\frac{2}{3}\right)^k \longrightarrow 0$ as $k \longrightarrow \infty$, our proposition follows readily. □

5.3. The Cantor Function and the Devil's Staircase

The content of this section will not be used henceforth except in Problem 9.18. It may be viewed as an "interesting aside."

Continuous surjective maps from Γ onto $\mathcal{Q}$, other than the one in (5.1.2), may be obtained in terms of the function $\gamma : \Gamma \longrightarrow \mathcal{I}$, which is defined by

$$\gamma(0_{\dot{3}}(2t_1)(2t_2)(2t_3)(2t_4)\ldots) = 0_{\dot{2}}t_1t_2t_3t_4\ldots, t_j = 0 \quad \text{or} \quad 1. \tag{5.3.1}$$

Note that this mapping is not injective because both $0_{\dot{3}}0\overline{2}$ and $0_{\dot{3}}2$, for example, are mapped into $0_{\dot{2}}1$.

It follows immediately from the definition in (5.3.1) that this function is surjective and continuous. Hence, $f_h \circ \gamma, f_p \circ \gamma$, *and* $f_s \circ \gamma$ *all map* Γ *continuously onto* $\mathcal{Q}$.

The definition of γ may be extended continuously into all of $\mathcal{I}$ by defining it on Γ^c as follows: If $t \in \Gamma^c$, then $t \in (a, b)$ where the open interval (a, b) is one of those that has been removed from $\mathcal{I}$ in the construction of the Cantor set. We have seen in the proof of Lemma 5.1 that

$$a = 0_{\dot{3}}(2a_1)(2a_2)(2a_3)\ldots(2a_{n-1})1 = 0_{\dot{3}}(2a_1)(2a_2)(2a_3)\ldots(2a_{n-1})0\overline{2},$$
$$b = 0_{\dot{3}}(2a_1)(2a_2)(2a_3)\ldots(2a_{n-1})2.$$

(See (5.2.1).) Hence, $\gamma(a) = 0_{\dot{2}}a_1a_2a_3\ldots a_{n-1}0\overline{1}$, $\gamma(b) = 0_{\dot{2}}a_1a_2a_3\ldots a_{n-1}1$, i.e., $\gamma(a) = \gamma(b)$. We define

$$\gamma(t) = \gamma(a)(= \gamma(b)), t \in (a, b) \tag{5.3.1a}$$

on each open interval (a, b) that has been removed in the construction of the Cantor set. The function

$$\gamma : \mathcal{I} \stackrel{\text{onto}}{\longrightarrow} \mathcal{I}$$

defined in (5.3.1) and (5.3.1a) is called the *Cantor function*. By construction, γ is increasing. *An increasing function has, at worst, jump discontinuities and is otherwise continuous.* To see this, we only need to show that $\lim_{t\to c-0} \gamma(t)$ and $\lim_{t\to c+0} \gamma(t)$ exist for all $c \in \mathcal{I}$ and that

$$\lim_{t\to c-0} \gamma(t) \leq \gamma(c) \leq \lim_{t\to c+0} \gamma(t).$$

(At $c = 0$ the left member and at $c = 1$ the right member of this inequality is to be omitted.)

Let $\mathcal{A} = \{\gamma(t) | 0 < t < c\}$. Since γ is increasing, $\mathcal{A}$ is bounded above by $\gamma(c)$. Hence $\sup \mathcal{A} = \alpha$ exists and $\alpha \leq \gamma(c)$. For any $\epsilon > 0$, there is a $\gamma(t_1) \in \mathcal{A}$ such that $\alpha - \epsilon < \gamma(t_1) \leq \alpha$. Since γ is increasing, $\alpha - \epsilon < \gamma(t) \leq \alpha$

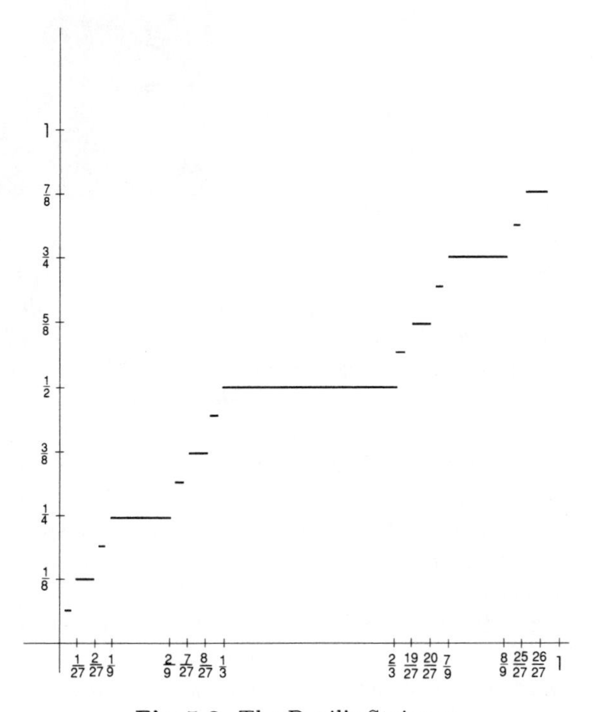

Fig. 5.3. The Devil's Staircase

for all $t \in (t_1, c)$. Therefore, $|\gamma(t) - \alpha| < \epsilon$ whenever $t_1 < t < c$, meaning that $\lim_{t \to c-0} \gamma(t)$ exists and $\lim_{t \to c-0} \gamma(t) = \alpha \leq \gamma(c)$. That $\lim_{t \to c+0} \gamma(t)$ exists and $\lim_{t \to c+0} \gamma(t) \geq \gamma(c)$ may be shown in the same manner.

But γ cannot have any jump discontinuities. If it did have one at c, then $\mathcal{I} = \gamma_*(\mathcal{I})$ would have a gap between $\gamma(c-0)$ and $\gamma(c+0)$ with $\gamma(c)$ in between. Hence, *the Cantor function is continuous.*

The graph of γ is called *the Devil's Staircase.* (See Fig. 5.3.)

5.4. Lebesgue's Definition of a Space-Filling Curve

Henri Leon Lebesgue (1875–1941) was born in Beauvais and died in Paris. He attended the École Normal Supérieur from 1894–1897 to qualify as a teacher of the mathematical sciences. After having taught at the Lycée of Nancy from 1899 to 1902, he received his doctor's degree in the sciences in 1902 with his thesis "Integrale, longeur, aire," which was published in the Annali di Matematica pura et applicata, and ushered in modern integration theory. Unfortunately, it also triggered an unpleasant controversy involving his teacher É. Borel ("Lebesgue et Borel s'opposèrent dans une violente polémique", Denjoy, Felix, and Montel [1]), who claimed that Lebesgue had not found anything new and that the family of Borel sets is identical with the family of (Lebesgue) measurable sets. This question was not settled until 1917, when M. Souslin constructed what he called analytic sets that are Lebesgue measurable but are not Borel sets. Lebesgue, who has 165 publications to his credit, had already published 82 books and papers by 1920, at which time he was appointed professor of applied analysis with geometry at the faculty of Sciences in Paris. In 1921, he became professor of mathematics at the Collége de France. In 1922, he was elected to the Academy of Sciences in Paris. A. Denjoy, L. Felix, and P. Montel called the progress immense when Lebesgue introduced the space of measurable functions. ("Lebesgue intrudisit l'espèce des fonctions mesurables. Les progrès était immense". Loc. Cit. p. 3). Paul Montel sums up Lebesgue's life with the words "He was a great scholar, an admirable professor, a man of incomparable intellectual high-mindedness. His influence on the development of mathematics will continue for a long time through his own works, as well as those he inspired" ("Il a été un grand savant, un professeur

admirable, un homme d'une incomparable noblesse morale. Son influence sur le développement des mathématiques continuera longtemps à s'exercer par ses oeuvres propres et par celles qu'il a inspirées." Denjoy, Felix, and Montel [1], p. 18) (For details see Hawkins [1], Denjoy [1], Denjoy, Felix, and Montel [1], Montel [1].)

In 1904, H. Lebesgue extended the mapping f in (5.1.2) continuously into $\mathcal{I}$ by linear interpolation (Lebesgue [1], [2]). If (a_n, b_n) is an interval that is removed in the construction of Γ at the nth step, then the extended function f_l is defined in terms of f on that interval as follows:

$$f_l(t) = \frac{1}{b_n - a_n}[f(a_n)(b_n - t) + f(b_n)(t - a_n)], a_n \leq t \leq b_n. \tag{5.4.1}$$

By construction, *f_l is continuous on Γ^c and maps $\mathcal{I}$* onto $\mathcal{Q}$ (because f has done that already). We will now demonstrate the following:

(5.4.1) Theorem. *The mapping f_l as defined in (5.1.2) and extended into Γ^c as in (5.4.1) is continuous on $\mathcal{I}$.*

Proof. If $t_0 \in \Gamma^c$, then f_l is continuous at t_0 by construction. If $t_0 \in \Gamma$, then, by Lemma 5.2.2, it is either a left accumulation point, a right accumulation point or an accumulation point. We will only consider the case where t_0 is a left accumulation point, namely $0_{\dot{3}}(2t_1)(2t_2)(2t_3)\ldots(2t_n)$, which makes it a right endpoint of an interval that has been removed in the construction of Γ. By Theorem 5.1, f (the restriction of f_l to Γ) is continuous on Γ, and, since Γ is compact, uniformly continuous on Γ. Hence, for any $\epsilon > 0$, there is a $\delta > 0$ such that

$$\|f(t_1) - f(t_2)\| < \epsilon \tag{5.4.2}$$

for all $t_1, t_2 \in \Gamma$ for which $|t_1 - t_2| < \delta$.

Let us now examine $\| f(t) - f(t_0) \|$. There are two possibilities: $t \in \Gamma$ or $t \in \Gamma^c$. If $t \in \Gamma$, then, by (5.4.2),

$$\|f(t) - f(t_0)\| < \epsilon \quad \text{for} \quad |t - t_0| < \delta, t \in \Gamma. \tag{5.4.3}$$

If $t \in \Gamma^c$, then, $t \in (a, b)$, where (a, b) is one of the intervals that have been removed in the construction of $\Gamma : a, b \in \Gamma, (a, b) \subset \Gamma^c$. From (5.4.1),

$$f_l(t) - f_l(t_0) = \frac{1}{b-a}[(f(b) - f(t_0))(t - a) + (f(a) - f(t_0))(b - t)]$$

for all $t \in (a, b)$. Now, let (a, b) be any such interval with $t_0 < a < b < t_0 + \delta$. In view of (5.4.3),

$$\|f_l(t) - f_l(t_0)\| < \frac{1}{b-a}(t - a + b - t)\epsilon = \epsilon$$

for all $t \in (a, b) \subset \Gamma^c$.

This, together with the fact that to the left of t_0 (t_0 being the right endpoint of a removed interval) f_l is continuous by construction, implies the continuity of f_l at $t_0 \in \Gamma$. An analogous proof applies to a right accumulation point of Γ, and both proofs together take care of a two-sided accumulation point of Γ. We may therefore consider our proposition as proved. (A simpler "geometric" proof will be given in the next section.) □

Lebesgue noted that one can obtain in an analogous manner a continuous mapping onto the n-dimensional unit cube, and even onto the $\aleph_0$-dimensional unit cube, having as a consequence that $\mathfrak{c}^{\aleph_0} = \mathfrak{c}$ (Lebesgue [2]). This will be discussed in greater detail in Section 6.9. A generalization of Lebesgue's construction may be found in Hahn [1], pp. 51–55.

Since f_l is defined by linear interpolation as in (5.4.1) on the countably many disjoint open intervals, the union of which is Γ^c, it is differentiable on Γ^c. Since the mapping $f : \Gamma \longrightarrow \mathcal{Q}$ as defined in (5.1.2), which is the restriction of f_l to Γ, is nowhere differentiable on Γ, it follows that f_l is nowhere differentiable on Γ. Finally, since $\Lambda_1(\Gamma) = 0$, we have the following:

(5.4.2) Theorem. *Lebesgue's space-filling curve is differentiable almost everywhere. Specifically, it is differentiable on Γ^c, and it is nowhere differentiable on Γ.*

Heretofore, all the known space-filling curves were nowhere differentiable. It was assumed that this was inevitable. However, Lebesgue's example put an end to that.

On the other hand, Lebesgue's space-filling curve lacks a property shared by all the curves that were discussed in Chapters 2, 3, and 4, namely, that every part of these curves is again a space-filling curve. For example, the portion of the Hilbert curve on the interval $[(k-1)/2^{2n}, k/2^{2n}]$, $n = 1, 2, 3, \ldots$, $k = 1, 2, 3, \ldots, 2^{2n}$, is an exact replica of the entire curve, reduced in the ratio $2^n : 1$. (The same is true for the Peano curve on the intervals $[(k-1)/3^{2n}, k/3^{2n}]$ with the reduction ratio $3^n : 1$ and the Sierpiński curve on $[(k-1)/2^n, k/2^n]$ with the reduction ratio $2^{n/2} : 1$.) What E. Cesàro said about the von Koch curve (Cesàro [2]) applies just as well to these curves: "Si elle était douée de vie, il ne serait pas possible de l'anéantir, sans la suprimer d'emblée car elle renaî̂trait sans cesse des profoundeurs de ses triangles, comme la vie dans l'universe." In reference to Hilbert's and Peano's curves, *triangles* will have to be replaced by *carrés*. ("If she were imbued with life,—in France, a curve is feminine—it would not be possible to destroy her without eliminating her altogether, because she will be reborn ceaselessly from the depth of her triangles, like life in the universe.") By contrast, were one to "kill" Lebesgue's curve, say on Γ_2, it would be truly "dead" because all that would be left are two segments of unit length that bisect each other at a right angle, a fitting grave marker.

Now that we know that there are almost everywhere differentiable space-filling curves, we may ask the question: can an everywhere differentiable curve be space-filling? In 1980, M. Morayne proved that, if $f = \begin{pmatrix} \varphi \\ \psi \end{pmatrix} : \mathbb{R} \longrightarrow \mathbb{R}^2$, where φ is Lebesgue-measurable, where at least one of $\varphi'(t), \psi'(t)$ exists for each $t \in \mathbb{R}$, and where $f_*(\mathbb{R})$ is Lebesgue-measurable, then $\Lambda_2(f_*(\mathbb{R})) = 0$ (Morayne [1], Theorem 3, p. 131). Suppose that $f_1 = \begin{pmatrix} \varphi_1 \\ \psi_1 \end{pmatrix} : \mathcal{I} \longrightarrow \mathbb{R}^2$ is differentiable and $J_2(f_{1*}(\mathcal{I})) > 0$, i.e., f_1 represents a differentiable space-filling curve. Let $\delta : \mathbb{R} \longleftrightarrow (0,1)$ represent a diffeomorphism. Note that we still have $J_2(f_{1*}((0,1))) > 0$ and that the function $f_1 \circ \delta : \mathbb{R} \longrightarrow \mathbb{R}^2$ satisfies the conditions of Morayne's theorem and, hence, $\Lambda_2(f_{1*} \circ \delta(\mathbb{R})) = \Lambda_2(f_{1*}((0,1))) = \Lambda_2(f_{1*}(\mathcal{I})) = J_2(f_{1*}(\mathcal{I})) = 0$, contrary to our assumption. Hence: *there is no differentiable space-filling curve.* (Morayne's paper had quite a remarkable publishing history: It was submitted in 1980 to the *Colloquium Mathematicum* and first published in 1984 in Volume XLVIII, fasc. 2 with a first page from somebody else's paper, rendering it incomprehensible. It was then republished in Volume XLIX, fasc. 1, squeezed in between pages 136 and 137 and, finally, published in its entirety and properly paginated in 1987 in Volume LIII, fasc. 1.) For generalizations of Morayne's results, see Morayne [2] and Cichoń and Morayne [1].

5.5. Approximating Polygons for the Lebesgue Curve

In a 1986 paper (Sagan [2]) we introduced one notion and in 1993 (Sagan [10]) another notion of approximating polygons for Lebesgue's space-filling curve. With a view towards Chapter 8, we will now adopt the latter. To arrive at this concept, let us first map the closed intervals $\left[0, \frac{1}{9}\right]$, $\left[\frac{2}{9}, \frac{1}{3}\right]$, $\left[\frac{2}{3}, \frac{7}{9}\right]$, and $\left[\frac{8}{9}, 1\right]$ that make up the set Γ_2 in (5.2.3) into the four subsquares $\mathcal{Q}_{00}, \mathcal{Q}_{01}, \mathcal{Q}_{10}, \mathcal{Q}_{11}$ of $\mathcal{Q}$ as in Fig. 5.5.1.

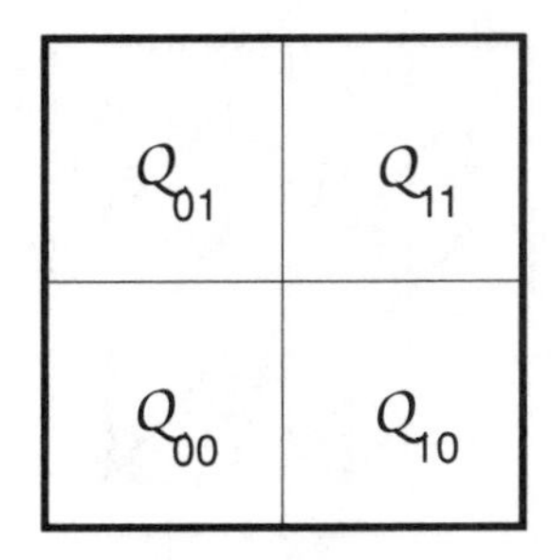

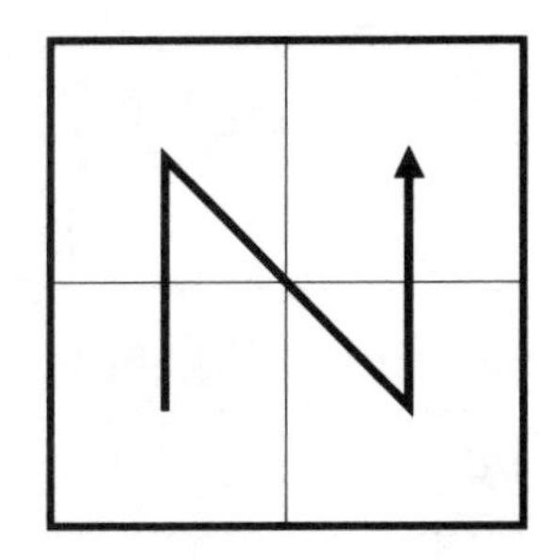

Fig. 5.5.1. Mapping Γ_2 into $\mathcal{Q}$

These subsquares are obtained from Q by means of the similarity transformations

$$\mathfrak{L}_{00}\begin{pmatrix}\xi\\ \eta\end{pmatrix} = \frac{1}{2}\begin{pmatrix}\xi\\ \eta\end{pmatrix} + \frac{1}{2}\begin{pmatrix}0\\ 0\end{pmatrix} \triangleq \frac{1}{2}x + \frac{1}{2}\ell_{00}$$

$$\mathfrak{L}_{01}\begin{pmatrix}\xi\\ \eta\end{pmatrix} = \frac{1}{2}\begin{pmatrix}\xi\\ \eta\end{pmatrix} + \frac{1}{2}\begin{pmatrix}0\\ 1\end{pmatrix} \triangleq \frac{1}{2}x + \frac{1}{2}\ell_{01}$$

$$\mathfrak{L}_{10}\begin{pmatrix}\xi\\ \eta\end{pmatrix} = \frac{1}{2}\begin{pmatrix}\xi\\ \eta\end{pmatrix} + \frac{1}{2}\begin{pmatrix}1\\ 0\end{pmatrix} \triangleq \frac{1}{2}x + \frac{1}{2}\ell_{10}$$

$$\mathfrak{L}_{11}\begin{pmatrix}\xi\\ \eta\end{pmatrix} = \frac{1}{2}\begin{pmatrix}\xi\\ \eta\end{pmatrix} + \frac{1}{2}\begin{pmatrix}1\\ 1\end{pmatrix} \triangleq \frac{1}{2}x + \frac{1}{2}\ell_{11}.$$

Since $t \in \left[0, \frac{1}{9}\right]$ when $t = 0_{\dot{3}}00(2t_3)(2t_4)\ldots, t \in \left[\frac{2}{9}, \frac{1}{3}\right]$ when $t = 0_{\dot{3}}02(2t_3)(2t_4)\ldots, t \in \left[\frac{2}{3}, \frac{7}{9}\right]$ when $t = 0_{\dot{3}}20(2t_3)(2t_4)\ldots$, and $t \in \left[\frac{8}{9}, 1\right]$ when $t = 0_{\dot{3}}22(2t_3)(2t_4)\ldots$, we have

$$\ell_{ij} = \begin{pmatrix}t_i\\ t_j\end{pmatrix}.$$

Next, we apply these transformations to the partitioned square in Fig. 5.5.1 to obtain the mapping of Γ_4 into the 16 subsquares of $\mathcal{Q}$, which are depicted in Fig. 5.5.2.

We keep repeating this process ad infinitum and obtain for the image $f(t)$ for $t \in \Gamma$, as in Sections 2.4, 3.4, and 4.3,

$$f(0_{\dot{3}}(2t_1)(2t_2)(2t_3)\ldots) = \sum_{j=1}^{\infty} \frac{1}{2^j}\ell_{2j-1\;2j}$$

$$= \sum_{j=1}^{\infty} \frac{1}{2^j}\begin{pmatrix}t_{2j-1}\\ t_{2j}\end{pmatrix} = \begin{pmatrix}0_{\dot{2}}t_1t_3t_5\ldots\\ 0_{\dot{2}}t_2t_4t_6\ldots\end{pmatrix},$$

which is precisely (5.1.2).

The beginning points of the closed intervals in Γ_{2n} have the form

$$t_b = 0_{\dot{3}}(2t_1)(2t_2)\ldots(2t_{2n})$$

$\mathcal{Q}_{0101}$	$\mathcal{Q}_{0111}$	$\mathcal{Q}_{1101}$	$\mathcal{Q}_{1111}$
$\mathcal{Q}_{0100}$	$\mathcal{Q}_{0110}$	$\mathcal{Q}_{1100}$	$\mathcal{Q}_{1110}$
$\mathcal{Q}_{0001}$	$\mathcal{Q}_{0011}$	$\mathcal{Q}_{1001}$	$\mathcal{Q}_{1011}$
$\mathcal{Q}_{0000}$	$\mathcal{Q}_{0010}$	$\mathcal{Q}_{1000}$	$\mathcal{Q}_{1010}$

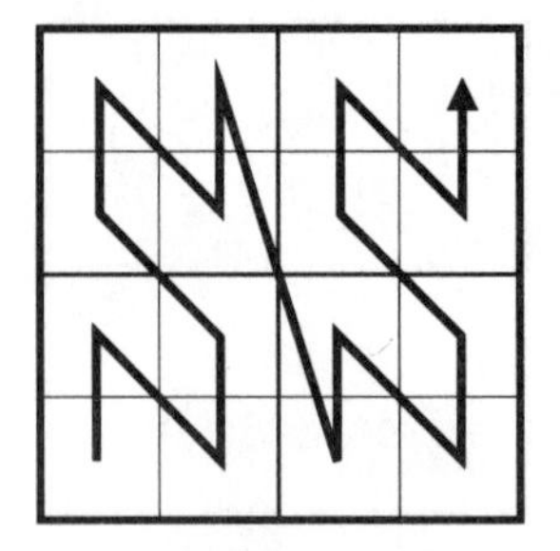

Fig. 5.5.2. Mapping of Γ_4 into $\mathcal{Q}$

and map into

$$f(t_b) = \begin{pmatrix} 0_{\dot{2}} t_1 t_3 \ldots t_{2n-1} \\ 0_{\dot{2}} t_2 t_4 \ldots t_{2n} \end{pmatrix},$$

and the endpoints have the form

$$t_e = t_b + \frac{1}{3^{2n}} = 0_{\dot{3}}(2t_1)(2t_2)\ldots(2t_{2n})\overline{2},$$

which map into

$$f(t_e) = \begin{pmatrix} 0_{\dot{2}} t_1 t_3 \ldots t_{2n-1}\overline{1} \\ 0_{\dot{2}} t_2 t_4 \ldots t_{2n}\overline{1} \end{pmatrix},$$

which is diagonally across from $f(t_b)$ in the subsquare $\mathcal{Q}_{t_1 t_2 \ldots t_{2n}}$ of side-length $\frac{1}{2^n}$. So, Lebesgue's space-filling curve enters the subsquare $\mathcal{Q}_{t_1 t_2 \ldots t_{2n}}$ at the left-lower corner and exits that square at the right-upper corner. We now join the exit point from one square to the entry point of the following square by a straight line, which is precisely the linear interpolation (5.4.1). (The open intervals of $\mathcal{I}\backslash\Gamma_{2n}$ are mapped linearly onto the connecting lines ("joins") from the right-upper corner of one subsquare to the left-lower corner of the next following subsquare.) This suggests the definition of the approximating polygons as the polygons that are put together from the off-diagonals of the subsquares and the joins. We have depicted the first two such polygons in Fig. 5.5.3, where we have rounded off some corners to avoid disguising the proper progression. These polygons are approximating polygons in the conventional sense because, within each subsquare, the distance from the approximating polygon to Lebesgue's curve is bounded above by the length $2^{-n+1/2}$ of the diagonal, and they coincide with Lebesgue's curve along the joins. Hence, they form a sequence of continuous curves that converges uniformly to the Lebesgue curve, which yields an alternate *proof for the continuity of Lebesgue's space-filling curve.*

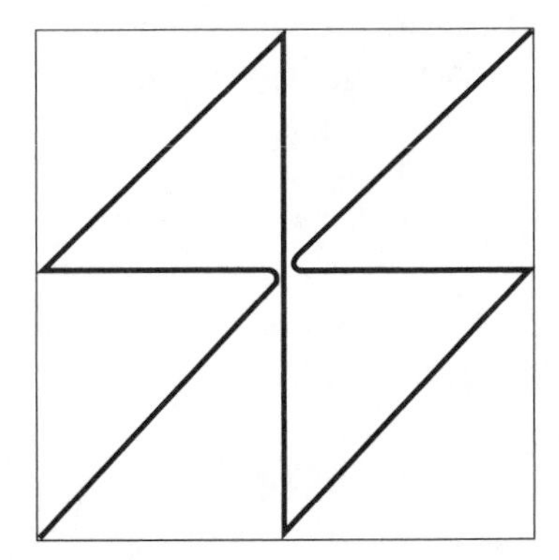
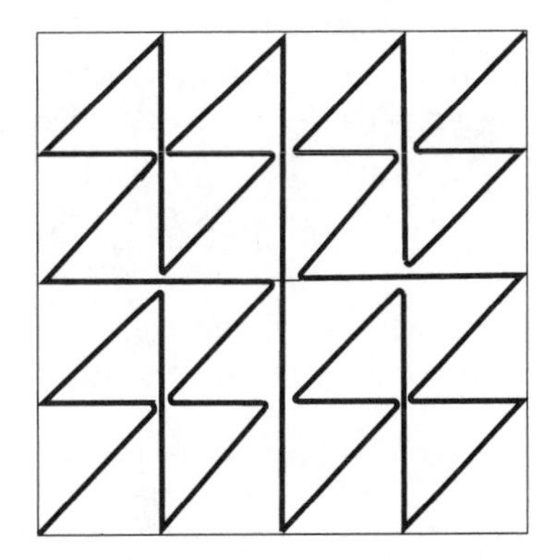

Fig. 5.5.3. First and Second Approximating Polygon for Lebesgue's Space-Filling Curve

5.6. Problems

1. Prove that the mapping defined in (5.1.3) is surjective, continuous, and nowhere differentiable.
2. Which of the following points lie in the Cantor set Γ: $0_{\dot{3}}02021\overline{2}, 0_{\dot{3}}010\overline{2}, 0_{\dot{3}}0\overline{1}, 0_{\dot{3}}10\overline{2}, 0_{\dot{3}}2221$?
3. Show that $1/4 \in \Gamma$.
4. Show: The set $\mathcal{S}$ of all beginning points and endpoints of all the intervals that have been removed in the construction of the Cantor set is dense in Γ, i.e., for any $\gamma \in \Gamma$ and any $\epsilon > 0$, there is an element $s \in \mathcal{S}$ such that $|s - \gamma| < \epsilon$.
5. Prove that Lebesgue's space-filling curve is continuous at all right accumulation points of Γ and all two-sided accumulation points of Γ.
6. Prove that the mapping γ from Γ to $\mathcal{I}$, as defined in (5.3.1), is continuous and surjective.
7. Show that $\gamma'(t) = 0$ a.e. where γ is the Cantor function that is defined in (5.3.1) and (5.3.1a).
8. Let $\mathcal{A} \subset \mathbb{R}$. $\mathcal{D}(\mathcal{A}) = \{x - y | x, y \in \mathcal{A}\}$ is called the *difference set of* $\mathcal{A}$. Show that $\mathcal{D}(\Gamma) = \mathcal{I}$.
9. Let $\alpha \in (0,1)$ and let Γ_α denote the set that is obtained from $\mathcal{I}$ by removing the open interval $\left(\frac{1}{2} - \frac{\alpha}{4}, \frac{1}{2} + \frac{\alpha}{4}\right)$ of length $\alpha/2$, and then removing from the remaining two closed intervals Γ_0, Γ_1 the middle open intervals of length $\alpha/8$ each, and then from the remaining four closed intervals $\Gamma_{00}, \Gamma_{01}, \Gamma_{10}, \Gamma_{11}$ the middle open intervals of length $\alpha/32$ each, etc.... Let $\Gamma_\alpha^{(1)} = \Gamma_0 \cup \Gamma_1, \Gamma_\alpha^{(2)} = \Gamma_{00} \cup \Gamma_{01} \cup \Gamma_{10} \cup \Gamma_{11}$, etc.... Then $\Gamma_\alpha = \bigcap_{j=1}^{\infty} \Gamma_\alpha^{(j)}$. One calls Γ_α a Cantor set of positive measure. Show that $\Lambda_1(\Gamma_\alpha) = 1 - \alpha$.
10. Find $\Lambda_2(\Gamma_\alpha \times \Gamma_\alpha)$ for $\alpha \in (0,1)$. (For the definition of Γ_α, see Problem 9.)
11. Find $f_\ell'(t)$ for $t \in \Omega_j$ for $j = 1,2,3,4$, where Ω_j is defined in (5.2.2).
12. A bounded subset $\mathcal{S}$ of $\mathbb{E}^n$ has n-dimensional Jordan content 0 if, for any $\epsilon > 0$, there are finitely many n-dimensional closed right parallelepipeds $\mathcal{C}_1, \mathcal{C}_2, \ldots, \mathcal{C}_n$ such that

$$\mathcal{S} \subseteq \bigcup_{k=1}^{n} \mathcal{C}_k \quad \text{and} \quad \sum_{k=1}^{n} \mathrm{Vol}(\mathcal{C}_k) < \epsilon.$$

$\mathcal{S}$ has Lebesgue measure 0 if one can find for any $\epsilon > 0$ a sequence of open right parallelepipeds $\mathcal{O}_1, \mathcal{O}_2, \ldots, \mathcal{O}_n, \ldots$ such that

$$\mathcal{S} \subseteq \bigcup_{k=1}^{\infty} \mathcal{O}_k \quad \text{and} \quad \sum_{k=1}^{\infty} \mathrm{Vol}(\mathcal{O}_k) < \epsilon.$$

Show: Every subset of $\mathbb{E}^n$ with Jordan content 0 has a Lebesgue measure 0.

13. Define an approximating polygon of nth order for Lebesgue's space-filling curve as the polygonal path that joins the image points of $0_{\dot{3}}(2t_1)(2t_2)(2t_3)\ldots(2t_n)$ in their natural order. Sketch the approximating polygons of order 1–7.
14. "Geometrize" the three-dimensional Lebesgue curve that is obtained by extending the mapping (5.1.3) continuously into $\mathcal{I}$, as was done in Section 4.5 for the two-dimensional curve, and sketch a first and second approximating polygon.

Chapter 6

Continuous Images of a Line Segment

6.1. Preliminary Remarks and a Global Characterization of Continuity

With squares and triangles and all their continuous images revealed as continuous images of the interval $\mathcal{I}$, the question arose as to the general characterization of such sets. In 1908, A. Schoenflies found such a characterization (see Schoenflies [1], p. 237), which, by its very nature, only applies to sets in the two-dimensional plane. In 1913, Hans Hahn and Stefan Mazurkiewicz independently arrived at a complete characterization of such sets in $\mathbb{E}^n$ (Hahn [2], [3], Mazurkiewicz [1], [2], [3]). Their results can be extended to apply to even more general spaces, as we will point out at the end of Section 6.8. We will follow Hahn's approach, commenting on Mazurkiewicz' work briefly in Section 6.7. The development of this chapter may be viewed as a generalization of Lebesgue's construction of a space-filling curve. Two elements made this construction possible: First, the square $\mathcal{Q}$ emerged as a continuous image of the Cantor set and, secondly, it was possible to extend the definition of the mapping continuously from Γ into $\mathcal{I}$ by linear interpolation, i.e., by joining the image of the left endpoint of an interval that has been removed in the construction of the Cantor set to the image of the right endpoint by a straight line (which lies in the square $\mathcal{Q}$). Hausdorff has shown that every compact set is a continuous image of the Cantor set. But compactness is not enough to ensure that the above mentioned images can be joined by a continuous arc that remains in the set in such a manner that the extended map is continuous. The conditions of connectedness and local connectedness, in addition to compactness, ensure that this can be done. Since these three conditions turn out to be necessary as well, a complete characterization of the continuous images of a line segment will be attained. Marie Torhorst, a student of Hahn at the University of Bonn, showed that local connectedness is necessary and sufficient for the boundary of a two-dimensional point set to be a continuous curve (Torhorst [1]). Hahn demonstrated on the basis of Torhorst's result that his condition, when applied to $\mathbb{E}^2$, and Schoenflies' condition, are equivalent

(Hahn [4]). We will comment on some other characterizations of continuous images of a line segment at the end of Section 6.8.

To discuss Hahn's result, we have to take a brief detour into point set topology. This, section and Sections 6.2, 6.3, and 6.5 will be devoted to a discourse on compactness, connectedness, local connectedness, and how these properties are affected by continuous transformations.

Hans Hahn (1879–1934) was born and died in Vienna. In 1898, he enrolled at the University of Vienna to study law. After one year, he switched to mathematics at the University of Straßburg. For the fall semester 1900–1901 he went to the University of Munich, and returned to the University of Vienna for the spring semester of 1901. He remained there and received his doctor's degree in July 1902. He had written a thesis on the theory of the second variation under the direction of Gustav, Ritter von Escherich. He spent another year at the University of Vienna to attend lectures by Mertens and Boltzman, and seminars by Boltzman, von Escherich, and Mertens. He spent the fall semester 1903–1904 in Göttingen to attend lectures by Hilbert and Minkowski, and to participate in seminars by Hilbert, Klein, and Minkowski. In the spring semester of 1904, he returned to Vienna and attended lectures and seminars given by Wirtinger. After the obligatory stint as a Privat Dozent at the University of Vienna from 1905 to 1909, he was appointed professor at the University of Czernowitz in the Bukowina, which was then part of the Austro-Hungarian Empire. Participating in World War I, he was severely wounded in 1915 and was decorated for gallantry. In 1916, he went to the University of Bonn, and returned to the University of Vienna in 1921. There he stayed until his untimely death, following cancer surgery. Hahn made many original contributions to the development of the Lagrange problem in the Calculus of variations. He introduced the term "anormal extremals," a distinction that was already recognized by Mayer and von Escherich in some earlier work, and he was the first to derive the Weierstraß condition for the Lagrange Problem. He also shares credit with St. Banach for the theorem on the extension of a linear functional from a linear manifold of a normed linear space into the entire space with preservation of the norm. In the theory of analytic functions, he generalized Weierstraß' factorization theorem to analytic functions of two variables. He also made

significant contributions to topology and the theory of functions of a real variable. (For more details, see Mayrhofer [1].)

Stefan Mazurkiewicz (1888–1945) was born in Warsaw and died nearby. He attended the Universities of Cracow, Munich, Göttingen, and L'vov and received his doctor's degree from the latter, under the guidance of W. Sierpiński in 1913, having written a thesis on space-filling curves ("O Krzywych wypelniajacych kwadrat"). In 1915, he became professor of mathematics at the newly created Polish University of Warsaw. He became one of the foremost representatives of Polish science. He was dean at the University of Warsaw for nine years and its pro-rector in 1937. From 1935, he served as the general secretary of the Warsaw Scientific Society. Together with W. Sierpiński, he was chief editor of the Fundamenta Mathematicae, which he, Sierpiński, and Z. Janiszewski founded in 1920. (Janiszewski died shortly thereafter.) His contributions to Mathematics fall mainly into the areas of topology, real analysis, theory of analytic functions, and probability theory. From 1940 to 1944, when all Polish institutions of higher learning were shut down by the occupying power, he, along with Sierpiński and hundreds of other scholars, continued to teach, at great personal risk to their safety, in Poland's "clandestine university." (It is estimated that from 1943 to 44, 3,800 students were enrolled in this underground university in Warsaw alone.) Like Hahn, he also suffered an untimely death, at the age of 57, at a time when he was absorbed with ambitious plans for the restructuring of Polish science. After the 1944 Warsaw uprising, the citizenry was evacuated by the occupying power to be put into camps. On the trek south, he contracted a stomach ailment as a result of a rich meal that was served to him by a well-meaning samaritan, who had taken pity on the emaciated scholar. With no possibility for curing his illness in war-torn Poland, he succumbed to it one year later. (The circumstances of Mazurkiewicz' death were related to the author by A. Schinzel on November 3, 1992. He, in turn, heard it from his former professor, W. Sierpiński). For more details, see Knaster [1], Kuratowski ([3], pp. 163–167), Mazurkiewicz ([6], pp. 9–26).

In the following, we will consider functions f, whose domain $\mathcal{D}(f)$ is a subset of $\mathbb{E}^m$ and whose range $\mathcal{R}(f)$ is a subset of $\mathbb{E}^n$. The independent variable, being a vector in $\mathbb{E}^m$, will be denoted by lowercase letters such as x, except when $m = 1$, when we will denote the independent variable by t.

(6.1.1) Definition. *$N_\delta(a) = \{x \in \mathbb{E}^n | \ \| x - a \| < \delta\}$ is called a δ-neighborhood of $a \in \mathbb{E}^n$ and $N'_\delta(a) = \{0 < \| x - a \| < \delta\}$ is called a deleted δ-neighborhood of $a \in \mathbb{E}^n$.*

(6.1.2) Definition. *$a \in \mathcal{A} \subseteq \mathbb{E}^n$ is called an interior point of $\mathcal{A}$ if there exists a $\delta > 0$ such that $N_\delta(a) \subset \mathcal{A}$. $\mathcal{A}^0$ denotes the set of all interior points of $\mathcal{A}$ and is called the interior of $\mathcal{A}$. $\mathcal{A}$ is open if it consists of interior points only, i.e., $\mathcal{A}^0 = \mathcal{A}$. $\mathcal{A}$ is closed if its complement $\mathcal{A}^c$ is open.*

(6.1.3) Definition. *$a \in \mathbb{E}^n$ is called an accumulation point of $\mathcal{A} \subseteq \mathbb{E}^n$ if for any $\delta > 0, N'_\delta(a) \cap \mathcal{A} \neq \emptyset$. The union of a set $\mathcal{A}$ with the set of all its accumulation points is called the closure of $\mathcal{A}$ and denoted by $\overline{\mathcal{A}}$. $\mathcal{A}$ is dense in $\mathcal{B}$ if $\overline{\mathcal{A}} = \mathcal{B}$.*

(6.1) Lemma. *A set is closed if and only if it contains all its accumulation points, i.e., $\overline{\mathcal{A}} = \mathcal{A}$.*

Proof. (a) Suppose that $\mathcal{A}$ is closed and a is an accumulation point but $a \notin \mathcal{A}$. Then, $a \in \mathcal{A}^c$. $\mathcal{A}^c$ is open, hence, there is a $\delta > 0$ such that $N_\delta(a) \subset \mathcal{A}^c$. But then a cannot be an accumulation point of $\mathcal{A}$.

(b) Suppose $\overline{\mathcal{A}} = \mathcal{A}$. If $x \notin \mathcal{A}$, then $x \notin \overline{\mathcal{A}}$ and there is a $\delta > 0$ such that $N_\delta(x)$ contains no point from $\mathcal{A}$, i.e., x is an interior point of $\mathcal{A}^c$. This is true for all $x \notin \mathcal{A}$ and $\mathcal{A}^c$ is open and, by Definition 6.1.2, $\mathcal{A}$ is closed. □

(6.1.4) Definition. *$f_*(\mathcal{A}) = \{f(x) \in \mathcal{R}(f) | x \in \mathcal{A} \cap \mathcal{D}(f)\}$ is called the direct image of $\mathcal{A}$ under f.*

(This is the same as Definition 1.3.1, which we restate here for completeness' sake.)

(6.1.5) Definition. *$f^*(\mathcal{B}) = \{x \in \mathcal{D}(f) | f(x) \in \mathcal{B} \cap \mathcal{R}(f)\}$ is called the inverse image of $\mathcal{B}$ under f.*

Note that

$$f^* f_*(\mathcal{A}) \supseteq \mathcal{A} \cap \mathcal{D}(f), f_* f^*(\mathcal{B}) = \mathcal{B} \cap \mathcal{R}(f) \tag{6.1.1}$$

$$f_*(\mathcal{A}_1 \cup \mathcal{A}_2) = f_*(\mathcal{A}_1) \cup f_*(\mathcal{A}_2), f^*(\mathcal{B}_1 \cup \mathcal{B}_2) = f^*(\mathcal{B}_1) \cup f^*(\mathcal{B}_2) \tag{6.1.2}$$

$$f_*(\mathcal{A}_1 \cap \mathcal{A}_2) \subseteq f_*(\mathcal{A}_1) \cap f_*(\mathcal{A}_2), f^*(\mathcal{B}_1 \cap \mathcal{B}_2) = f^*(\mathcal{B}_1) \cap f^*(\mathcal{B}_2) \tag{6.1.3}$$

$$f_*(\mathcal{A}) = \emptyset \Rightarrow \mathcal{A} \cap \mathcal{D}(f) = \emptyset, f^*(\mathcal{B}) = \emptyset \Rightarrow \mathcal{B} \cap \mathcal{R}(f) = \emptyset \tag{6.1.4}$$

$$(f \circ g)^*(\mathcal{A}) = g^* \circ f^*(\mathcal{A}). \tag{6.1.5}$$

The latter may be seen as follows: $t \in (f \circ g)^*(\mathcal{A})$ implies $f(g(t)) \in \mathcal{A}$, which implies, in turn, that $g(t) \in f^*(\mathcal{A})$ and, hence, $t \in g^*(f^*(\mathcal{A}))$. On the other hand, $t \in g^*(f^*(\mathcal{A}))$ means that $g(t) \in f^*(\mathcal{A})$ and, hence, $f(g(t)) \in f_*(f^*(\mathcal{A})) = \mathcal{A} \cap R(f)$. Therefore, $t \in (f \circ g)^*(\mathcal{A})$.

(6.1.6) Definition. $f : \mathcal{D}(f) \longrightarrow \mathbb{E}^n, \mathcal{D}(f) \subset \mathbb{E}^m$, *is continuous at* $a \in \mathcal{D}(f)$ *if for all* $\epsilon > 0$, *there is a* $\delta(\epsilon, a) > 0$ *such that*

$$f_*(N_{\delta(\epsilon,a)}(a)) \subset N_\epsilon(f(a)).$$

A function is continuous on $\mathcal{A} \subseteq \mathcal{D}(f)$ *if it is continuous at every point of* $\mathcal{A}$.

The following global characterization of continuity will prove helpful in the subsequent developments:

(6.1.1) Theorem. $f : \mathcal{D}(f) \longrightarrow \mathbb{E}^n$ *is continuous in* $\mathcal{D}(f)$ *if and only if for every open set* $\Omega \subseteq \mathbb{E}^n$ *there is an open set* $\Omega_1 \subseteq \mathbb{E}^m$ *such that* $f^*(\Omega) = \Omega_1 \cap \mathcal{D}(f)$.

Proof. (a) We assume that the condition in Definition 6.1.6 is satisfied. Let $\Omega \subseteq \mathbb{E}^n$ represent an open set. If $f^*(\Omega) = \emptyset$, we are through. So, we may assume that $f^*(\Omega) \neq \emptyset$. It is our objective to construct an open set Ω_1 that satisfies the condition of the theorem. If $a \in f^*(\Omega)$, then $f(a) \in \Omega$. Since Ω is open, there is an $\epsilon > 0$ so that $N_\epsilon(f(a)) \subseteq \Omega$. By Definition 6.1.6, there is a $\delta(\epsilon, a) > 0$ so that $f_*(N_{\delta(\epsilon,a)}(a)) \subset N_\epsilon(f(a))$. Hence, $f_*(N_{\delta(\epsilon,a)}(a)) \subset \Omega$ and, by (6.1.1),

$$N_{\delta(\epsilon,a)}(a) \cap \mathcal{D}(f) \subseteq f^*(\Omega). \tag{6.1.6}$$

We construct such a $N_{\delta(\epsilon,a)}(a)$ for every $a \in f^*(\Omega)$ and denote

$$N_{\delta(\epsilon,a)}(a) = \mathcal{U}(a).$$

(Note that ϵ depends on a.)

Let

$$\Omega_1 = \bigcup_{a \in f^*(\Omega)} \mathcal{U}(a).$$

By construction, Ω_1 is open. If $x \in f^*(\Omega)$, then $x \in \Omega_1$. Hence, $f^*(\Omega) \subseteq \Omega_1$. Since we also have, from the definition of the inverse image, that $f^*(\Omega) \subseteq \mathcal{D}(f)$, it follows that

$$f^*(\Omega) \subseteq \Omega_1 \cap \mathcal{D}(f). \tag{6.1.7}$$

On the other hand, if $x \in \Omega_1 \cap \mathcal{D}(f)$, then $x \in \mathcal{U}(a)$ for some $a \in f^*(\Omega)$, and, hence, by (6.1.6), $x \in f^*(\Omega)$, i.e.,

$$\Omega_1 \cap \mathcal{D}(f) \subseteq f^*(\Omega). \tag{6.1.8}$$

Together, (6.1.7) and (6.1.8), yield the desired result.

(b) We assume that the condition of the theorem is met. Let $a \in \mathcal{D}(f)$. For every $\epsilon > 0, N_\epsilon(f(a)) = \Omega$ is an open set in $\mathbb{E}^n$ and $a \in f^*(N_\epsilon(f(a)))$.

By hypothesis, there is an open set $\Omega_1 \subseteq \mathbb{E}^m$ such that $f^*(N_\epsilon(f(a))) = \Omega_1 \cap \mathcal{D}(f)$. Since $a \in \Omega_1$ and sine Ω_1 is open, there is a $\delta(\epsilon, a) > 0$ such that $N_{\delta(\epsilon,a)}(a) \subseteq \Omega_1$. Since $N_{\delta(\epsilon,a)}(a) \cap \mathcal{D}(f) \subseteq \Omega_1 \cap \mathcal{D}(f) = f^*(N_\epsilon(f(a))$, the condition of Definition 6.1.6 follows directly. □

We need the following equivalent global characterization of continuity for our proof of Netto's theorem in Section 6.4:

(6.1.2) Theorem. *$f : \mathcal{D}(f) \longrightarrow \mathbb{E}^n$ is continuous on $\mathcal{D}(f)$ if and only if for every closed set $\mathcal{K} \subseteq \mathbb{E}^n$ there is a closed set $\mathcal{K}_1 \subseteq \mathbb{E}^m$ such that $f^*(\mathcal{K}) = \mathcal{K}_1 \cap \mathcal{D}(f)$.*

(In effect, this characterization is formally obtained from the one in Theorem 6.1.1 by the interchange of "open" with "closed" wherever it occurs.)

Proof. Let $\mathcal{A} \subseteq \mathbb{E}^n$. From (6.1.2),

$$f^*(\mathbb{E}^n) = f^*(\mathcal{A} \cup \mathcal{A}^c) = f^*(\mathcal{A}) \cup f^*(\mathcal{A}^c) = \mathcal{D}(f)$$

and we have

$$\mathcal{D}(f) = f^*(\mathcal{A}) \cup f^*(\mathcal{A}^c). \tag{6.1.9}$$

Since $\mathcal{A} \cap \mathcal{A}^c = \emptyset$, we have from (6.1.3)

$$\emptyset = f^*(\emptyset) = f^*(\mathcal{A} \cap \mathcal{A}^c) = f^*(\mathcal{A}) \cap f^*(\mathcal{A}^c)$$

and, hence,

$$f^*(\mathcal{A}) \cap f^*(\mathcal{A}^c) = \emptyset. \tag{6.1.10}$$

If $\mathcal{A}_1 \subseteq \mathbb{E}^n$ is such that $\mathcal{A}_1 \cap \mathcal{D}(f) = f^*(\mathcal{A})$, then

$$\mathcal{A}_1^c \cap f^*(\mathcal{A}) = \emptyset \tag{6.1.11}$$

and

$$\mathcal{D}(f) = (\mathcal{A}_1 \cap \mathcal{D}(f)) \cup (\mathcal{A}_1^c \cap \mathcal{D}(f)) = f^*(\mathcal{A}) \cup (\mathcal{A}_1^c \cap \mathcal{D}(f)). \tag{6.1.12}$$

From (6.1.9) and (6.1.12)

$$f^*(\mathcal{A}) \cup f^*(\mathcal{A}^c) = f^*(\mathcal{A}) \cup (\mathcal{A}_1^c \cap \mathcal{D}(f)).$$

$f^*(\mathcal{A})$ has no elements in common with $f^*(\mathcal{A}^c)$ by (6.1.10) and has no elements in common with $\mathcal{A}_1^c$ by (6.1.11)). Hence,

$$f^*(\mathcal{A}) = \mathcal{A}_1 \cap \mathcal{D}(f) \tag{6.1.13}$$

implies

$$f^*(\mathcal{A}^c) = \mathcal{A}_1^c \cap \mathcal{D}(f). \tag{6.1.14}$$

We are now ready to prove Theorem 6.1.2:

(a) Let $\mathcal{A}^c$ denote a closed set. Then, $\mathcal{A}$ is open and, by Theorem 6.1.1, there is an open set $\mathcal{A}_1$ such that $f^*(\mathcal{A}) = \mathcal{A}_1 \cap \mathcal{D}(f)$, which is (6.1.13). This implies (6.1.14), namely, $f^*(\mathcal{A}^c) = \mathcal{A}_1^c \cap \mathcal{D}(f)$, where $\mathcal{A}_1^c$ is closed and we see that Theorem 6.1.1 implies Theorem 6.1.2.

(b) Let $\mathcal{A}^c$ denote an open set. Then, $\mathcal{A}$ is closed and, by Theorem 6.1.2, there is a closed set $\mathcal{A}_1$ such that $f^*(\mathcal{A}) = \mathcal{A}_1 \cap \mathcal{D}(f)$, which implies $f^*(\mathcal{A}^c) = \mathcal{A}_1^c \cap \mathcal{D}(f)$, where $\mathcal{A}^c$ is open, and we see that Theorem 6.1.2 implies Theorem 6.1.1. □

6.2. Compact Sets

There are many ways of defining a compact set. For our purposes, we find it practical to define compactness in terms of the Heine-Borel property:

(6.2.1) Definition. *A set $\mathcal{S} \subset \mathbb{E}^n$ has the Heine-Borel property if every open cover of $\mathcal{S}$, meaning every collection of open sets $\{\Omega_\alpha | \alpha \in A\}$, for some index set A, so that $\mathcal{S} \subseteq \bigcup_{\alpha \in A} \Omega_\alpha$, contains a finite subcover*

$$\{\Omega_{\alpha_1}, \Omega_{\alpha_2}, \ldots, \Omega_{\alpha_k}\} \tag{6.2.1}$$

with $\mathcal{S} \subseteq \bigcup_{j=1}^{k} \Omega_{\alpha_j}$, where $\alpha_j \in A$.

(6.2.2) Definition. *A set $\mathcal{K} \subset \mathbb{E}^n$ is called compact if it has the Heine-Borel property.*

Next to the Heine-Borel property, we also will need another characterization of compact sets. For that purpose, we need the concept of a *bounded set.*

(6.2.3) Definition. *A set $\mathcal{S} \subset \mathbb{E}^n$ is called bounded if there is an $R > 0$ such that $\| x \| \leq R$ for all $x \in \mathcal{S}$.*

(6.2.1) Theorem. *(Heine-Borel Theorem):$\mathcal{K} \subset \mathbb{E}^n$ is compact if and only if it is bounded and closed.*

Proof. (a) We assume that $\mathcal{K}$ has the Heine-Borel property. Clearly, $\{N_k(0) | k = 1, 2, 3, \ldots\}$ is an open cover of $\mathcal{K}$. There is a finite subcover $\{N_{k_1}(0), N_{k_2}(0), \ldots, N_{k_j}(0)\}$. Let $k_0 = \max\{k_1, k_2, \ldots, k_j\}$. Then $\mathcal{K} \subseteq N_{k_0}(0) = \{x \in \mathbb{E}^n | \; \| x \| < k_0\}$ and $\mathcal{K}$ is bounded. (See also Fig. 6.2.1(a).)

Suppose $\mathcal{K}$ is not closed, i.e., there is an accumulation point a of $\mathcal{K}$ which is not in $\mathcal{K}$. Then, $\{\Omega_k | k = 1, 2, \ldots\}$ with $\Omega_k = \{x \in \mathbb{E}^n | \; \| x - a \| > \frac{1}{k}, k = 1, 2, 3 \ldots\}$ is an open cover of $\mathcal{K}$. By hypothesis, a finite subcover $\{\Omega_k | k = k_1, k_2, \ldots, k_j\}$ will do. If $k_0 = \max\{k_1, k_2, \ldots, k_j\}$, then $\mathcal{K} \subseteq \Omega_{k_0}$

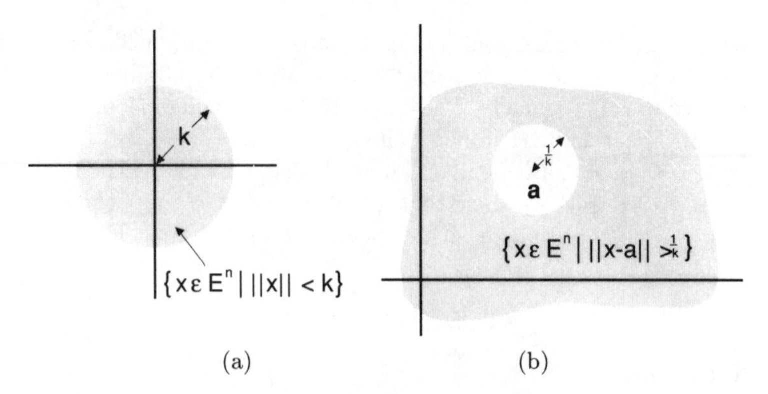

Fig. 6.2.1. Illustration of the Proof of the Heine-Borel Theorem

and $\{x \in \mathbb{E}^n | \parallel x - a \parallel \leq \frac{1}{k_0})$ contains no points from $\mathcal{K}$, i.e., a cannot be an accumulation point of $\mathcal{K}$. Hence, $\mathcal{K}$ contains all its accumulation points and is, therefore, closed. (See also Fig. 6.2.1(b).)

(b) We assume that $\mathcal{K}$ is bounded and closed. Since it is bounded, it is contained in a closed n-dimensional cube $\mathcal{W}$ of sidelength s for some $s > 0$. If $\Omega = \{\Omega_\alpha | \alpha \in A\}$ is an open cover of $\mathcal{K}$, then $\{\Omega, \mathcal{K}^c\}$ is an open cover of $\mathcal{W}$ (because $\mathcal{K}^c$ is open, $\mathcal{K}$ being closed, and $\{\Omega, \mathcal{K}^c\}$ covers the entire space). In order to show that $\mathcal{K}$ has the Heine-Borel property, we only need to show that $\mathcal{W}$ has it. Suppose it does not. We partition $\mathcal{W}$ into 2^n congruent subcubes of sidelength $s/2$ each. At least one, say $\mathcal{W}_1$, does not have the Heine-Borel property. We partition $\mathcal{W}_1$ into 2^n congruent subcubes of sidelength $s/4$ each and at least one, say $\mathcal{W}_2$, does not have the Heine-Borel property. We continue this process and arrive at a sequence of nested closed cubes $\mathcal{W} \supset \mathcal{W}_1 \supset \mathcal{W}_2 \supset \mathcal{W}_3 \supset \ldots$, with $\mathcal{W}_k$ having sidelength $s/2^k$, which, by Cantor's intersection theorem, defines a unique point $p \in \mathcal{W}$. Since $\{\Omega, \mathcal{K}^c)$ is an open cover of $\mathcal{W}, p \in \Omega_{\alpha_0}$ for some $\alpha_0 \in A$ or $p \in \mathcal{K}^c$. In either case, there is a $\delta > 0$ such that $N_\delta(p) \subseteq \Omega_{\alpha_0}$ or $N_\delta(p) \subseteq \mathcal{K}^c$. But then, all cubes from the sequence $\{\mathcal{W}_k\}$, from a certain subscript on, lie in $N_\delta(p) \subseteq \Omega_{\alpha_0}$ or in $N_\delta(p) \subseteq \mathcal{K}^c$, contrary to our assumption that none of them has the Heine-Borel property. □

It is now a relatively simple matter to show that compactness is invariant under continuous transformations:

(6.2.2) Theorem. *If $f : \mathcal{K} \longrightarrow \mathbb{E}^n$ is continuous and $\mathcal{K}$ is compact, then $f_*(\mathcal{K})$ is compact.*

Proof. Let $\{\Omega_\alpha | \alpha \in A\}$ represent an open cover of $f_*(\mathcal{K})$. Since f is continuous, we have from Theorem 6.1.1 that there are open sets $\Omega_{1\alpha}, \alpha \in A$, such that $f^*(\Omega_\alpha) = \Omega_{1\alpha} \cap \mathcal{K}$. If $x \in \mathcal{K}$, then $f(x) \in \Omega_\alpha$ for some $\alpha \in A$. Hence, $x \in f^*(\Omega_\alpha)$. Hence, $\{f^*(\Omega_\alpha) | \alpha \in A\}$ is an open cover of $\mathcal{K}$.

Since $\mathcal{K}$ is compact, $\{\Omega_{1\alpha_1}, \Omega_{1\alpha_2}, \ldots \Omega_{1\alpha_j}\}$ will cover $\mathcal{K}$. For every $f(x) \in f_*(\mathcal{K}), x \in \Omega_{1\alpha_i}$ for some $i \in \{1, 2, 3, \ldots, j\}$ and, hence, $f(x) \in \Omega_{\alpha_i}$. Thus, $\{\Omega_{\alpha_1}, \Omega_{\alpha_2}, \ldots, \Omega_{\alpha_j}\}$ covers $f_*(\mathcal{K})$ and $f_*(\mathcal{K})$ stands revealed as compact.

□

Corollary to Theorem 6.2.2: *If $f : \mathcal{I} \longrightarrow \mathbb{E}^n$ is continuous, then $f_*(\mathcal{I})$ is compact.*

Proof. $\mathcal{I}$ is bounded and closed and, hence, by Theorem 6.2.1, compact. □

The following result will be needed in Section 6.7, and again in Chapter 8.

(6.2.1) Lemma. *If $\mathcal{D}$ is everywhere dense in $\mathcal{I}$, $\mathcal{M}$ is compact, and $g : \mathcal{D} \longrightarrow \mathcal{M}$ is uniformly continuous, then there exists a unique continuous extension of g into $\mathcal{I}$, i.e., there is a unique continuous function $f : \mathcal{I} \longrightarrow \mathcal{M}$ such that $f(t) = g(t)$ for all $t \in \mathcal{D}$.*

Proof. Since $\overline{\mathcal{D}} = \mathcal{I}$, we can pick for each $t \in \mathcal{I}$ a sequence $\{t_n\} \longrightarrow t, t_n \in \mathcal{D}$, and we have, from the uniform continuity of g on $\mathcal{D}$, that $\{g(t_n)\}$ is a Cauchy sequence: $\| g(t_n) - g(t_m) \| < \epsilon/3$ for $|t_n - t_m| < \rho$ for some $\rho > 0$ which is independent of t_n, t_m. This is the case if $m, n > N(\epsilon)$ for some $N(\epsilon)$. Hence, $\lim_{n \to \infty} g(t_n)$ exists, and we define

$$f(t) = \lim_{n \to \infty} g(t_n), t \in \mathcal{I} \backslash \mathcal{D}$$

(this definition is independent of the choice of sequence that tends to t) and

$$f(t) = g(t), t \in \mathcal{D}.$$

This function f is uniformly continuous on $\mathcal{I}$: Let $t', t'' \in \mathcal{I}$, where we assume w.l.o.g. that $t' < t''$, so that $|t' - t''| < \rho$ and pick sequences $\{t'_n\} \searrow t', \{t''_n\} \nearrow t''$ with $t'_n, t''_n \in \mathcal{D}$. Then,

$$\| f(t') - f(t'') \| \leq \| f(t') - f(t'_n) \| + \| f(t'_n) - f(t''_m) \| + \| f(t''_m) - f(t'') \| .$$

By construction, $|t'_n - t''_m| < \rho$, which makes the middle term less than $\epsilon/3$. Since $\{f(t'_n)\} \longrightarrow f(t')$ and $\{f(t''_n)\} \longrightarrow f(t'')$, we can choose m, n so large that the first and last terms are each less than $\epsilon/3$ and we have $\| f(t') - f(t'') \| < \epsilon$ for all $|t' - t''| < \rho$, i.e., f is uniformly continuous on $\mathcal{I}$. Furthermore, since $\mathcal{M}$ is compact, $f(t) \in \mathcal{M}$ for all $t \in \mathcal{I}$. This function f is unique, because a continuous function that vanishes on $\mathcal{D}$ is identically 0 on $\mathcal{I}$: If $h(t) = 0$ for all $t \in \mathcal{D}$ and $t_0 \in \mathcal{I}$, then $h(t_n) = 0$ for any sequence $\{t_n\} \longrightarrow t_0$ with $t_n \in \mathcal{D}$, and, hence, $h(t_0) = \lim_{n \to \infty} h(t_n) = 0$. □

The following result will be needed in Sections 6.9 and 6.10, and again in Chapter 8:

(6.2.2) Lemma. *If $\mathcal{D}$ is dense in the compact set $\mathcal{K}$ and if $\mathcal{D}$ is compact, then $\mathcal{D} = \mathcal{K}$.*

Proof. $\overline{\mathcal{D}} = \mathcal{K}$ and $\overline{\mathcal{D}} = \mathcal{D}$. □

6.3. Connected Sets

We consider a set to be "hanging together" (or "connected") if it cannot be decomposed into two disjoint non-empty sets. More precisely:

(6.3) Definition. *A set $\mathcal{S} \subseteq \mathbb{E}^n$ is connected if one cannot find two open sets $\mathcal{A}, \mathcal{B} \subset \mathbb{E}^n$ such that*

$$\mathcal{A} \cap \mathcal{S} \neq \emptyset, \mathcal{B} \cap \mathcal{S} \neq \emptyset$$
$$(\mathcal{A} \cap \mathcal{S}) \cup (\mathcal{B} \cap \mathcal{S}) = \mathcal{S}$$
$$(\mathcal{A} \cap \mathcal{S}) \cap (\mathcal{B} \cap \mathcal{S}) = \emptyset.$$

If there are two such sets, we call them a disconnection of $\mathcal{S}$, and $\mathcal{S}$ disconnected.

The set in Fig. 6.3.1 which consists of all points of the closed interval $[-1, 1]$ on the y-axis and the polygonal path that joins the points $(1, 1)$, $(\frac{1}{2}, -1)$, $(\frac{1}{3}, 1)$, $(\frac{1}{4}, -1)$, $(\frac{1}{5}, 1) \ldots$, is connected, but the set $S = [-1, 0) \cup (0, 1]$ in Fig. 6.3.2 is not.

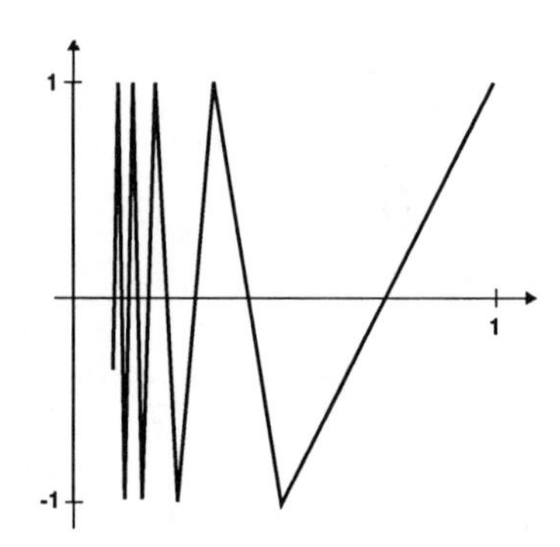

Fig. 6.3.1. Example of a Connected Set

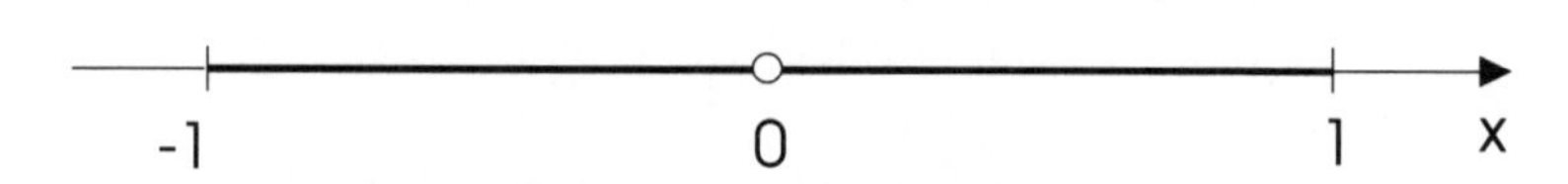

Fig. 6.3.2. Example of a Disconnected Set

Connectedness is invariant under continuous transformations:

(6.3.1) Theorem. *If $\mathcal{D}(f) \subseteq \mathbb{E}^m$ is connected and $f : \mathcal{D}(f) \longrightarrow \mathbb{E}^n$ is continuous, then $f_*(\mathcal{D}(f))$ is connected.*

Proof. (by contradiction): Suppose that $f_*(\mathcal{D}(f))$ is not connected. Then, there exists a disconnection $\mathcal{A}$, $\mathcal{B}$ such that

$$\mathcal{A} \cap f_*(\mathcal{D}(f)) \neq \emptyset, \mathcal{B} \cap f_*(\mathcal{D}(f)) \neq \emptyset \tag{6.3.1}$$

$$[\mathcal{A} \cap f_*(\mathcal{D}(f))] \cup [\mathcal{B} \cap f_*(\mathcal{D}(f))] = f_*(\mathcal{D}(f)) \tag{6.3.2}$$

$$[\mathcal{A} \cap f_*(\mathcal{D}(f))] \cap [\mathcal{B} \cap f_*(\mathcal{D}(f))] = \emptyset. \tag{6.3.3}$$

Since f is continuous, we have, from Theorem 6.1.1, two open sets $\mathcal{A}_1$, $\mathcal{B}_1$ such that

$$f^*(\mathcal{A}) = \mathcal{A}_1 \cap \mathcal{D}(f), f^*(\mathcal{B}) = \mathcal{B}_1 \cap \mathcal{D}(f). \tag{6.3.4}$$

We will demonstrate that $\mathcal{A}_1$, $\mathcal{B}_1$ are a disconnection of $\mathcal{D}(f)$, which is contrary to our hypothesis. From (6.3.2),

$$\begin{aligned} f^* f_*(\mathcal{D}(f)) &= f^*(\mathcal{R}(f)) = \mathcal{D}(f) = f^*[\mathcal{A} \cap f_*(\mathcal{D}(f)) \cup \mathcal{B} \cap f_*(\mathcal{D}(f))] \\ &= f^*[\mathcal{A} \cap f_*(\mathcal{D}(f))] \cup f^*[\mathcal{B} \cap f_*(\mathcal{D}(f))] = f^*(\mathcal{A}) \cup f^*(\mathcal{B}) \\ &= (\mathcal{A}_1 \cap \mathcal{D}(f)) \cup (\mathcal{B}_1 \cap \mathcal{D}(f)). \end{aligned} \tag{6.3.5}$$

From (6.3.1) and (6.1.4),

$$f^*(\mathcal{A}) \neq \emptyset, f^*(\mathcal{B}) \neq \emptyset.$$

Hence, from (6.3.4)

$$\mathcal{A}_1 \cap \mathcal{D}(f) \neq \emptyset, \mathcal{B}_1 \cap \mathcal{D}(f) \neq \emptyset. \tag{6.3.6}$$

Finally,

$$(\mathcal{A}_1 \cap \mathcal{D}(f)) \cap (\mathcal{B}_1 \cap \mathcal{D}(f)) = f^*(\mathcal{A}) \cap f^*(\mathcal{B}) = \emptyset, \tag{6.3.7}$$

since $\mathcal{A} \cap \mathcal{B} \cap f_*(\mathcal{D}(f)) = \emptyset$. Together, (6.3.5), (6.3.6), and (6.37) establish $\mathcal{A}_1$, $\mathcal{B}_1$ as a disconnection of $\mathcal{D}(f)$, which, by hypothesis, is connected. □

The only connected subsets of $\mathbb{R}$ are the intervals:

(6.3.2) Theorem. *A subset of $\mathbb{R}$ is connected if and only if it is an interval.*

Proof. (a) An interval is connected. An interval $\mathcal{I}$ has the following property: With any two distinct points $a, b \in \mathcal{I}$ with $a < b, [a, b] \subset \mathcal{I}$. (The empty set—as a subset of $\mathbb{R}$—and the set containing just one point are trivial connected subsets of $\mathbb{R}$). Let us assume that $\mathcal{I}$ is disconnected, i.e., there are two open subsets $\mathcal{A}$, $\mathcal{B} \subset \mathbb{R}$ that satisfy the three conditions of Definition 6.3 with $\mathcal{I}$ instead of $\mathcal{S}$. We may assume without loss of generality that $a \in \mathcal{A} \cap \mathcal{I}$, $b \subset \mathcal{B} \cap \mathcal{I}$.

Let

$$\mathcal{T} = \{t \in [a, b] | \, t \in \mathcal{A} \cap \mathcal{I}\}.$$

$\mathcal{T} \neq \emptyset$ because $a \in \mathcal{T}$. $\mathcal{T}$ is bounded above and has a least upper bound

$$t_0 = \sup \mathcal{T} \in [a, b]. \tag{6.3.8}$$

Suppose that $t_0 \in \mathcal{B} \cap \mathcal{I}$. $\mathcal{B}$ is open, hence $(t_0 - \delta, t_0] \subseteq \mathcal{B} \cap \mathcal{I}$ for some $\delta > 0$ ($\mathcal{I}$ contains all points between a and b!). In view of (6.3.8), there is a $t \in \mathcal{A} \cap \mathcal{I}, a \leq t \leq b$, such that $t_0 - \delta < t \leq t_0$. But this cannot be since $(\mathcal{A} \cap \mathcal{I}) \cap (\mathcal{B} \cap \mathcal{I}) = \emptyset$. Hence, $t_0 \notin \mathcal{B} \cap \mathcal{I}$. Suppose now that $t_0 \in \mathcal{A} \cap \mathcal{I}$. Since $\mathcal{A}$ is open and $t_0 < b$, we have $[t_0, t_0 + \delta) \subset \mathcal{A} \cap \mathcal{I}$ for some $\delta > 0$ which contradicts the definition of t_0 in (6.3.8). So, we arrive at the conclusion that $a, b \in \mathcal{I}, a < t_0 < b$, and $t_0 \notin (\mathcal{A} \cap \mathcal{I}) \cup (\mathcal{B} \cap \mathcal{I}) = \mathcal{I}$. Hence, $\mathcal{I}$ is not an interval, contrary to our assumption. (See also Fig. 6.3.3.)

(b) A connected subset of $\mathbb{R}$ is an interval: Suppose $\mathcal{I}$ it is not an interval but connected. Then, there are two points $a, b \in \mathcal{I}$ with $a < b$ and a point $z \notin \mathcal{I}$ such that $a < z < b$. Let

$$\mathcal{A} = \{x \in \mathbb{R} | x < z\}$$
$$\mathcal{B} = \{x \in \mathbb{R} | x > z\}.$$

Since $a \in \mathcal{A}, b \in \mathcal{B}$, neither set is empty and, contrary to our hypothesis they establish a disconnection of $\mathcal{I}$. $\square$

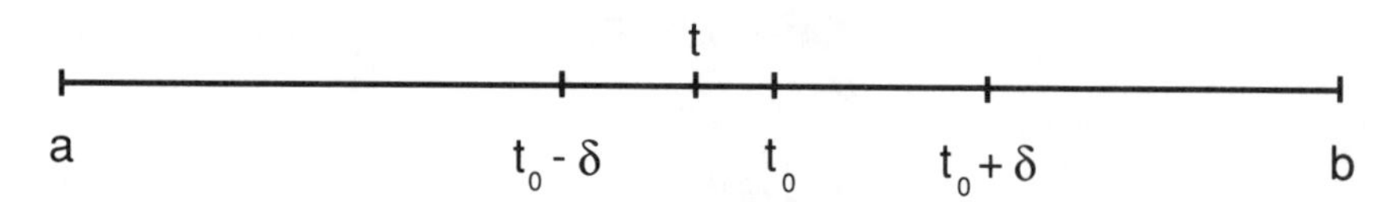

Fig. 6.3.3. Illustration for the Proof of Theorem 6.3.2

Corollary to Theorems 6.3.1 and 6.3.2. *If $f : \mathcal{I} \longrightarrow \mathbb{E}^n$ is continuous, then $f_*(\mathcal{I})$ is connected. If $n = 1$, then $f_*(\mathcal{I})$ is a closed interval.*

The following property of connected sets will be needed in Section 6.7:

(6.3) Lemma. *If $\mathcal{C}$ is closed and connected, then, for any two points $x', x'' \in \mathcal{C}$ and any $\eta > 0$, there exists an η-chain from x' to x'', i..e, a finite number of points $x_1, x_2, \ldots, x_n \in \mathcal{C}$ with $x' = x_1, x'' = x_n$, such that $\| x_{i+1} - x_i \| < \eta, i = 1, 2, 3, \ldots, n-1$.*

Proof. Choose $\eta > 0$ and let

$$\mathcal{A} = \{x \in \mathcal{C} | \text{there is an } \eta \text{ -chain from } x' \text{ to } x\}.$$

We want to show that $\mathcal{A} = \mathcal{C}$. We do this by contradiction. Let $\Omega = \bigcup_{a \in \mathcal{A}} N_\eta(a)$. If $x \in \mathcal{C} \cap \Omega$, then $x \in N_\eta(a) \cap \mathcal{C}$ for some $a \in \mathcal{A}$. Hence, there is an η-chain from x' to x via a. Hence, $x \in \mathcal{A}$, i.e., $\mathcal{C} \cap \Omega \subseteq \mathcal{A}$. Conversely, if $x \in \mathcal{A}$, then $x \in N_\eta(a) \subset \Omega$ and since $\mathcal{A} \subseteq \mathcal{C}$, we have $\mathcal{A} \subseteq \mathcal{C} \cap \Omega$. Thus, $\mathcal{A} = \mathcal{C} \cap \Omega$. Clearly, $\mathcal{C} \cap \Omega \neq \emptyset$ since $x' \in \mathcal{A}$. Let x denote an accumulation point of $\mathcal{A}$. Since $\mathcal{A} \subseteq \mathcal{C}$, x is also an accumulation point of $\mathcal{C}$ and since $\mathcal{C}$ is closed, $x \in \mathcal{C}$. Since $N'_\eta(x)$ contains a point from $\mathcal{A}$, say a, and since a is within η of $x \in \mathcal{C}$, there is an η-chain from x' to x via a and hence, $x \in \mathcal{A}$, meaning that $\mathcal{A}$ is closed and A^c is open. Since $\mathcal{A} = \mathcal{C} \cap \Omega$, we have $\mathcal{C} \cap \Omega \cap \mathcal{A}^c = \emptyset$. If $\mathcal{A}^c \cap \mathcal{C} \neq \emptyset$, then Ω and A^c are a disconnection of $\mathcal{C}$ because

$$A^c \cap \mathcal{C} \neq \emptyset, \Omega \cap \mathcal{C} \neq \emptyset$$
$$(A^c \cap \mathcal{C}) \cup (\Omega \cap \mathcal{C}) = \mathcal{C}$$
$$(A^c \cap \mathcal{C}) \cap (\Omega \cap \mathcal{C}) = \emptyset$$

and we have a contradiction. Hence, $\mathcal{A}^c \cap \mathcal{C} = \emptyset$, i.e., $\mathcal{C}$ has no points that are not in $\mathcal{A}$. Since $\mathcal{A} \subseteq \mathcal{C}$ by the definition of $\mathcal{A}$, we conclude that $\mathcal{A} = \mathcal{C}$, meaning that there is an η-chain from x' to all points of $\mathcal{C}$ and, consequently, there is an η-chain between any two points of $\mathcal{C}$ via x', and possibly others. □

(Note: Lemma 6.3 remains true if the assumption that $\mathcal{C}$ is closed is omitted. The reader is encouraged to prove it under these weaker conditions.)

6.4. Proof of Netto's Theorem

Now that we have the necessary tools at our disposal, we will give a proof of Netto's theorem, which we quoted in Section 1.3.

(6.4) Theorem. *(Netto): Any bijective map $f : \mathcal{I} \leftrightarrow \mathcal{Q}$ (or $\mathcal{W}$, or $\mathcal{T}$) is necessarily discontinuous.*

Proof. Since f is a bijective map onto $\mathcal{Q}$ (or $\mathcal{W}$, or $\mathcal{T}$), its inverse $f^{-1} : \mathcal{Q}$ (or $\mathcal{W}$, or $\mathcal{T}$) $\leftrightarrow \mathcal{I}$ exists. Let $g \triangleq f^{-1}$. Let $\mathcal{A} \subseteq \mathbb{R}$ denote a closed set. Then, $\mathcal{A} \cap \mathcal{I}$ is bounded and closed and, hence, compact. By Theorem 6.2.2, $f_*(\mathcal{A} \cap \mathcal{I})$ is compact and, hence, closed. Let $f_*(\mathcal{A} \cap \mathcal{I}) = \mathcal{A}_1$. Since $f_*(\mathcal{A} \cap \mathcal{I}) = g^*(\mathcal{A})$, we have $g^*(\mathcal{A}) = \mathcal{A}_1$ and since $\mathcal{A}_1 \subseteq f_*(\mathcal{I}) = \mathcal{D}(g)$, we can write $g^*(\mathcal{A}) = \mathcal{A}_1 \cap \mathcal{D}(g)$ and we have, from Theorem 6.1.2, that $g = f^{-1}$ is continuous.

Since f^{-1} is continuous, it maps, by Theorem 6.3.1, connected sets onto connected sets. We remove a point t_0 from the open interval $(0, 1)$ and its image $f(t_0)$ from $\mathcal{Q}$ (or $\mathcal{W}$, or $\mathcal{T}$). $\mathcal{Q} \backslash f(t_0)$ (or $\mathcal{W} \backslash f(t_0)$, or $\mathcal{T} \backslash f(t_0)$) are still connected but $\mathcal{I} \backslash t_0 = [0, t_0) \cup (t_0, 1]$ is not, and the continuous function f^{-1} appears to map a connected set onto a disconnected set. This is a contradiction to Theorem 6.3.1 and it follows that f cannot be continuous. □

Clearly, this proof not only applies to $\mathcal{Q}$, $\mathcal{W}$, and $\mathcal{T}$ but to any compact and connected set that contains at least two points and cannot be disconnected by the removal of one point. (For the more general case, see Greenberg [1].)

6.5. Locally Connected Sets

We can see from the corollary to Theorem 6.2.2 and the corollary to Theorems 6.3.1 and 6.3.2 that, for a set to be the continuous image of $\mathcal{I}$, it is compact and connected. This does not suffice, however. For example, the set in Fig. 6.3.1, which is compact and connected, cannot be obtained as a continuous image of $\mathcal{I}$. In 1913, Hans Hahn announced at a meeting of German Scientists and Physicians held in Vienna that he found a third condition that a set must satisfy in addition to compactness and connectedness in order to be the continuous image of $\mathcal{I}$ and that these three conditions together are also necessary. His lecture only dealt with the necessity of the condition (Hahn [2]). He published the part pertaining to the sufficiency in the Proceedings of the Imperial and Royal Academy of Science in Vienna in 1914 (Hahn [3]). The property in question is "local connectedness" or "Zusammenhang im Kleinen," as Hahn called it:

(6.5) Definition. *A set $\mathcal{M} \subseteq \mathbb{E}^n$ is locally connected if for every point $p \in \mathcal{M}$ and every $\epsilon > 0$ there is an $\eta(\epsilon, p) > 0$ such that for every point $p' \in N_{\eta(\epsilon,p)}(p) \cap \mathcal{M}$ there is a compact and connected subset $\mathcal{M}' \subseteq N_\epsilon(p)$ that contains p and p'.*

Other equivalent definitions can be found in the literature (Hausdorff [1], pp. 177, 179, or Willard [1], p. 199). To prevent possible confusion, let us point out that Willard defines "connected im kleinen" at a point of $\mathcal{M}$, as does Hahn, and then proceeds to show that a set is locally connected by his definition (Willard [1], p. 199) if it is "connected im kleinen" in every one of its points, i.e., connected "im kleinen" by Hahn's definition. (The difference is that Hahn uses the same term for the point-property as well as the set-property, while Willard uses "connected im kleinen" only for the point-property and uses "locally connected" for the set-property.)

A locally connected set need not be connected. The set in Fig. 6.3.2 is locally connected. On the other hand, a connected set need not be locally connected, as in the connected set in Fig. 6.3.1, which is not locally connected at any of the points in the vertical interval $[-1, 1]$ on the y-axis.

Local connectedness is not invariant under continuous transformations. It is possible to map a locally connected set continuously onto a set that is not locally connected. Take, for example, the set in Fig. 6.3.2 and define the following map from that set onto the set in Fig. 6.3.1: map the interval $[-1, -\frac{1}{3}]$ linearly onto the vertical interval $[-1, 1]$ with the point $(-1, 0)$

being mapped into the point $(0,-1)$, the interval $\left[-\frac{1}{3},0\right)$ linearly onto the vertical interval $(0,1]$ with the point $\left(-\frac{1}{3},0\right)$ being mapped into the point $(0,1)$, and the interval $(0,1]$ (by vertical projection) bijectively and continuously onto the zig-zag part of the set in Fig. 6.3.1. This mapping is continuous.

While local connectedness is not invariant under continuous transformations, local connectedness in conjunction with compactness is. The continuous image of a compact and locally connected set is compact and locally connected (Willard [1], p. 203). We do not need this result in such generality. It will suffice for our purpose to prove:

(6.5) Theorem. *If $f : \mathcal{I} \to \mathbb{E}^n$ is continuous, then $f_*(\mathcal{I})$ is locally connected. (It is also compact by Theorem 6.2.2 since $\mathcal{I}$ is compact).*

Proof. (by contradiction): We assume that $f_*(\mathcal{I}) = \mathcal{M}$ is not locally connected: There is a point $p \in \mathcal{M}$ and an $\epsilon > 0$ so that for all $\eta > 0$ there is a point $p' \in N_\eta(p) \cap \mathcal{M}$ such that every compact and connected subset of $\mathcal{M}$ containing p and p' has points outside $N_\epsilon(p)$.

Consider the sequence $\{\eta_j\}$ with $\eta_j > 0$ and $\lim_{j\to\infty} \eta_j = 0$. Then, for each η_j there is a point $p'_j \in N_{\eta_j(p)} \cap \mathcal{M}$ that does not have the required property. By construction, $\{p'_j\} \to p$. Since $\mathcal{M} = f_*(\mathcal{I})$, we have for each j at least one $t'_j \in \mathcal{I}$ such that $f(t'_j) = p'_j$. Since $\mathcal{I}$ is compact, there is a convergent subsequence $\{t'_{n_j}\} \to t' \in \mathcal{I}$. Since $f(t'_{n_j}) = p'_{n_j}$ and since f is continuous, we have $f(t') = p$. The line segment $< t'_{n_j}, t' >$ (which is the closed interval $[t'_{n_j}, t']$ if $t'_{n_j} < t'$, and the closed interval $[t', t'_{n_j}]$ if $t' < t'_{n_j}$) is mapped, by the corollaries to Theorems 6.2.2, and 6.3.1 and 6.3.2, onto a compact and connected subset $\mathcal{M}' \subseteq \mathcal{M}$ that contains p as well as p'_{n_j}. Since f is continuous, there is a $\delta > 0$ such that

$$\| f(t) - f(t') \| < \epsilon \quad \text{for all} \quad t \in N_\delta(t') \cap \mathcal{I}.$$

Since $\{t'_{n_j}\} \to t'$, there is a j_0 such that $t'_{n_j} \in N_\delta(t') \cap \mathcal{I}$ for all $j > j_0$. Hence, $f_*(< t'_{n_j}, t' >) \subset N_\epsilon(p)$, for all $j > j_0$, i.e., there is a compact and connected subset M' of M that lies in $N_\epsilon(p)$ and contains p, as well as p'_{n_j}, contrary to our assumption that p'_{n_j} does not have that property. □

6.6. A Theorem by Hausdorff

We see from the corollaries to Theorems 6.2.2, 6.3.1, 6.3.2, and Theorem 6.5, that a continuous image of $\mathcal{I}$ (or any closed line segment) is compact, connected, and locally connected. Hahn proved in 1913 that these conditions are also sufficient. His proof is laborious and not particularly lucid. In fact, it goes on and on for 35 pages! In 1927, there appeared the second edition of Felix Hausdorff's *Mengenlehre*, which contained a theorem asserting that each compact set is a continuous image of the Cantor set. (Hausdorff [1], pp. 154, 226). The same year, Alexandroff [1] published the same result. In 1928, Hahn used this theorem to give a proof that surpassed not only his original proof and the proof by Stefan Mazurkiewicz, but, according to Hahn (Hahn [7], p. 218), all other unpublished proofs, in terms of its simplicity and lucidity. We will present Hahn's second proof in Section 6.8. First, we have to prove Hausdorff's theorem and establish some additional pertinent properties of compact, connected, and locally connected sets.

(6.6) Theorem. *(Hausdorff): Every compact set is a continuous image of the Cantor set.*

Proof. (by construction):

Let $\mathcal{K}$ denote a compact set. We cover it by N_1-neighborhoods. Since $\mathcal{K}$ has the Heine-Borel property, there exists a finite subcover:

$$N_1(x_{i_1}), i_1 = 0, 1, 2, \ldots, 2^{n_1} - 1$$

(where we may assume, without loss of generality, that this subcover has 2^{n_1} members for some integer n_1—by counting some, if necessary, several times). Let

$$\mathcal{K}_{i_1} = \overline{N}_1(x_{i_1}) \cap \mathcal{K},$$

where $\overline{N}_1$ denotes the closure of $N_1 : \overline{N}_1(x_{i_1}) = \{x \in \mathbb{E}^n | \parallel x - x_{i_1} \parallel \leq 1\}$. We have $\bigcup_{i_1} \mathcal{K}_{i_1} = \mathcal{K}$. Next, we cover each $\mathcal{K}_{i_1}$ by $N_{1/2}$-neighborhoods and pick finite subcovers $N_{1/2}(x_{i_1 i_2})$, $i_2 = 0, 1, 2, \ldots, 2^{n_2} - 1$ (where we assume, without loss of generality, that each subcover has the same number of 2^{n_2} members). Let

$$\mathcal{K}_{i_1 i_2} = \overline{N}_{1/2}(x_{i_1 i_2}) \cap \mathcal{K}_{i_1}.$$

We have $\bigcup_{i_2} \mathcal{K}_{i_1 i_2} = \mathcal{K}_{i_1}$. We continue in this manner with $N_{1/4}-$ neighborhoods, $N_{1/8}$-neighborhoods, etc... and obtain nested sequences of compact sets

$$\mathcal{K} \supseteq \mathcal{K}_{i_1} \supseteq \mathcal{K}_{i_1 i_2} \supseteq \mathcal{K}_{i_1 i_2 i_3} \supseteq \ldots, \tag{6.6.1}$$

where $\mathcal{K}_{i_1 i_2 i_3 \ldots i_k}$ is contained in the closure of a $1/2^{k-1}$-neighborhood.

We may write any $t \in \Gamma$ as

$$t = 0_3(2b_1)(2b_2)\ldots(2b_{n_1})(2b_{n_1+1})(2b_{n_1+2})\ldots(2b_{n_1+n_2})(2b_{n_1+n_2+1})\cdots$$

(where $b_j = 0$ or 1 and where infinitely many trailing b_j might well be 0 and define a mapping $f : \Gamma \to \mathcal{K}$ by

$$f(t) = \mathcal{K}_{i_1} \cap \mathcal{K}_{i_1 i_2} \cap \mathcal{K}_{i_1 i_2 i_3} \cap \mathcal{K}_{i_1 i_2 i_3 i_4} \cap \ldots, \tag{6.6.2}$$

where $i_1 = (b_1 \ldots b_{n_1})_2$, $i_2 = (b_{n_1+1} \ldots b_{n_1+n_2})_2$, $i_3 = (b_{n_1+n_2+1} \cdots b_{n_1+n_2+n_3})_2, \cdots - (b_{n_1+\cdots+n_k+1} \ldots b_{n_1+\cdots+n_{k+1}})_2$ denotes the binary representation of the number i_{k+1}. By Cantor's intersection theorem, (6.6.2) associates with each $t \in \Gamma$ a unique point in $\mathcal{K}$.

This mapping is surjective because each point in $\mathcal{K}$ lies in at least one nested sequence (6.6.1) which corresponds to a binary sequence and which, in turn, corresponds to a point in Γ.

The mapping is also continuous: If $|t' - t''| < 1/3^{n+1}$, then

$$t' = 0_{\dot{3}}(2b_1)(2b_2)\ldots(2b_n)(2b_{n+1})(2b_{n+2})\ldots$$
$$t'' = 0_{\dot{3}}(2b_1)(2b_2)\ldots(2b_n)(2\beta_{n+1})(2\beta_{n+2})\ldots .$$

If $n_1 + n_2 + \cdots + n_j \leq n < n_1 + n_2 + \cdots + n_j + n_{j+1}$, then $f(t'), f(t'') \in \mathcal{K}_{i_1 i_2 \ldots i_j} \subseteq \overline{N}_{1/2^{j-1}}(x_{i_1 \ldots i_j})$, i.e., $\| f(t') - f(t'') \| < 1/2^{j-2}$. □

We see that Hausdorff's theorem is a straightforward generalization of the continuous mapping (6.1.2) or (6.1.3) from Γ onto $\mathcal{Q}$ or $\mathcal{W}$. There, we could extend the mapping continuously into the entire interval $\mathcal{I}$ by joining the images of the beginning points and endpoints of the open intervals that make up Γ^c. This was possible because the target set was convex. What we need is some guarantee that these points can still be connected by a continuous path that lies in $\mathcal{K}$, even when $\mathcal{K}$ is not convex. When $\mathcal{K}$ is compact, connected, and locally connected, then it is *uniformly pathwise connected* and such a construction is possible. Hahn [3] has shown this and we will present his proof in the next section. (Incidentally, this result is needed for Hahn's first proof as well as his second proof of the sufficiency of compactness, connectedness, and local connectedness in order for a set to be the continuous image of a line segment.)

6.7. Pathwise Connectedness

(6.7) Definition. *A set $\mathcal{A}$ is called pathwise connected if any two points $p', p'' \in \mathcal{A}$ can be joined by a continuous path that lies in $\mathcal{A}$, i.e., there is a continuous function $f : \mathcal{I} \to \mathcal{A}$ such that $f(0) = p', f(1) = p''$.*

(6.7) Lemma. *If $\mathcal{M}$ is compact, connected, and locally connected, then there is for all $\epsilon > 0$ an $\eta(\epsilon) > 0$ such that whenever two points $p', p'' \in \mathcal{M}$ are within $\eta(\epsilon)$ of each other, $\| p' - p'' \| < \eta(\epsilon)$, there is a compact, connected subset $\mathcal{M}' \subseteq \mathcal{M}$ which contains p', p'', and lies in an ϵ-neighborhood of p' as well as p'' : $p', p'' \in \mathcal{M}', \mathcal{M}' \subseteq N_\epsilon(p'), \mathcal{M}' \subseteq N_\epsilon(p'')$. (In words: p', p'' are connected by a compact, connected set that is close to p', p''. Note that $\eta \leq \epsilon$ since $p' \in \mathcal{M}'$ is within ϵ of p'' and vice versa.)*

Proof. (by contradiction): Suppose this is not true: There is an $\epsilon > 0$ so that for all $\eta_i > 0, i = 1, 2, 3, \ldots$, with $\lim_{i \to \infty} \eta_i = 0$, there are pairs of points $\{p'_i, p''_i\}$ within η_i of each other, $\| p'_i - p''_i \| < \eta_i$, so that any compact connection $\mathcal{M}_i'$ of p_i', p_i'' has points that are outside $N_\epsilon(p_i')$ or $N_\epsilon(p_i'')$. Since $\mathcal{M}$ is compact, there is a subsequence of $\{p_i', p_i''\}$ so that the subsequence of $\{p_i'\}$ converges to some $p' \in \mathcal{M}$, and a subsequence of that subsequence so that, in addition, the corresponding subsequence of $\{p_i''\}$ converges to some $p'' \in \mathcal{M}$. We call this subsequence of a subsequence $\{p_k', p_k''\}$. We have, by construction, $\lim_{k\to\infty} p_k' = p'$, $\lim_{k\to\infty} p_k'' = p''$. Since $\| p_k' - p_k'' \| < \eta_k$ and $\{\eta_k\} \to 0$, we have $p' = p'' \triangleq p$. Since $\mathcal{M}$ is closed, $p \in \mathcal{M}$. Since $\mathcal{M}$ is locally connected, there is for any $\epsilon/2 > 0$ an $\eta > 0$ such that, whenever $\overline{p} \in N_\eta(p)$, then $p, \overline{p}$ are connected by a compact subset $\mathcal{M}' \subseteq \mathcal{M}$ that lies entirely in an $\epsilon/2$-neighborhood of p. We choose k_0 so large that $p_k' \in N_\eta(p), p_k'' \in N_\eta(p)$ for all $k > k_0$. Then, p_k' is connected to p by a compact subset $\mathcal{M}' \subseteq \mathcal{M}$ and p_k'' is connected to p by a compact subset $\mathcal{M}'' \subseteq \mathcal{M}$ and both $\mathcal{M}', \mathcal{M}''$ lie in an $\epsilon/2$-neighborhood of p. $\mathcal{M} \cup \mathcal{M}'$ is compact and, since $\mathcal{M}$ and $\mathcal{M}'$ contain p, connected and, finally, by construction, $\mathcal{M}' \cup \mathcal{M}''$ lies in an ϵ-neighborhood of p_k', as well as in an ϵ-neighborhood of p_k'', contrary to our assumption. □

Since the η, the existence of which is asserted in Lemma 6.7, is independent of p and only depends on ϵ, we say that a compact, connected, and locally connected set is *uniformly locally connected.*

(6.7.1) Theorem. *A compact, connected, and locally connected set M is pathwise connected.*

Note: this is neither the case for the set in Fig. 6.3.1, which is not locally connected, nor for the set in Fig. 6.3.2, which is not connected. Also, observe that a compact and pathwise connected set need not be locally connected, as one may see from Fig. 6.7.1, where the angle that is formed by a spoke and its neighbor in the clockwise direction is half the angle formed by the spoke with its neighbor in the counterclockwise direction.

Proof. By Lemma 6.7, $\mathcal{M}$ is uniformly locally connected: For any $\epsilon_k > 0$, there is an $\eta_k > 0$ such that any two points $q_k', q_k'' \in \mathcal{M}$ that are within η_k of each other are connected by a compact set $\mathcal{M}_k \subseteq \mathcal{M}$ that lies in an ϵ_k-neighborhood of q_k' as well as in an ϵ_k-neighborhood of q_k''. Let us choose a sequence $\{\epsilon_k\}$ with $\epsilon_k > 0$ so that $\sum_{k=1}^{\infty} \epsilon_k < \infty$. By Lemma 6.3, there is an η_1-chain from p' to p'' and we may assume w.l.o.g. that there are $2^{n_1} + 1$ points in this chain, including the two endpoints, for some integer n_1 (if need be, we can always add points or count points several times):

$$p' = p_0^{(1)}, p_1^{(1)}, \ldots, p_{2^{n_1}}^{(1)} = p''. \tag{6.7.1}$$

By Lemma 6.7, each two adjacent points in this chain are connected by a compact set $\mathcal{M}_i^{(1)} \subseteq \mathcal{M}$ $(i = 1, 2, 3, \ldots, 2^{n_1})$, which lies in an

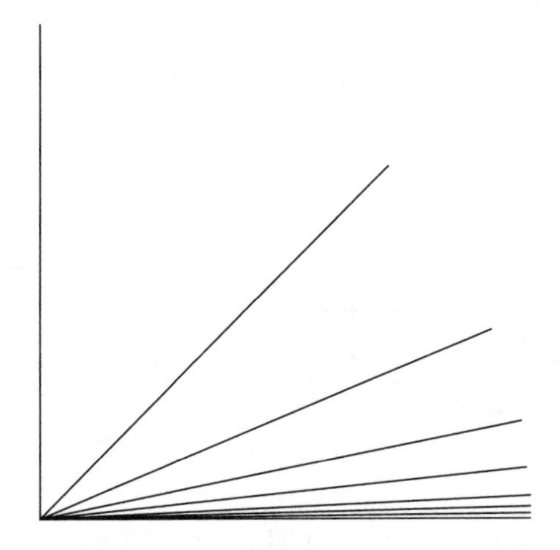

Fig. 6.7.1. Example of a Compact Pathwise Connected Set that is not Locally Connected

ϵ_1-neighborhood of either point. Hence, by Lemma, 6.3, there is an η_2 chain between any two adjacent points in (6.7.1) that lies in $\mathcal{M}_i^{(1)}$, where we may assume w.l.o.g. that each such chain consists of the same number $2^{n_2}+1$ points, for some suitable integer n_2:

$$p_i^{(1)} = p_{i.2^{n_2}}^{(2)}, p_{i.2^{n_2}+1}^{(2)}, \ldots, p_{(i+1).2^{n_2}}^{(2)} = p_{i+1}^{(1)}, \tag{6.7.2}$$

$i = 0, 1, 2, \ldots, 2^{n_1} - 1$.

Each two adjacent points in (6.7.2) are connected by a compact set $\mathcal{M}_j^{(2)} \subseteq \mathcal{M}_i^{(1)}$ that lies in an ϵ_2-neighborhood of the one point as well as in an ϵ_2-neighborhood of the other point. Again, we invoke Lemma 6.3 and construct η_3-chains between each two adjacent points of (6.7.2) that lie in $\mathcal{M}_j^{(2)}$ and have $2^{n_3}+1$ points each, etc.... (See also Fig. 6.7.2.)

After k such steps we arrive at a chain

$$p' = p_0^{(k)}, p_1^{(k)}, \ldots, p_{2^{n_1+n_2+\cdots+n_k}}^{(k)} = p'', \tag{6.7.3),}$$

where each two adjacent points are connected by some compact set $\mathcal{M}_r^{(k)} \subseteq \mathcal{M}_j^{(k-1)}$, which lies in an ϵ_k-neighborhood of either one.

As a first step towards constructing a continuous function $f : \mathcal{I} \to \mathcal{M}$ we define a function g for all $t \in \mathcal{I}$ that can be represented in the form

$$t_{i,k} = \frac{i}{2^{n_1+n_2+\cdots+n_k}}, i = 0, 1, 2, \ldots, 2^{n_1+n_2+\cdots+n_k}$$

as follows:

$$g(t_{i,k}) = p_i^{(k)}, i = 0, 1, 2, \ldots, 2^{n_1+n_2+\cdots+n_k}. \tag{6.7.4}$$

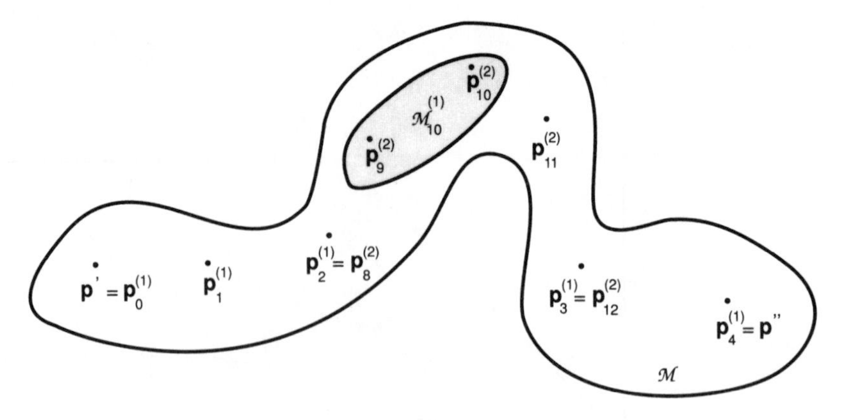

Fig. 6.7.2. Illustration of the Proof of Theorem 6.7.2 for $n_1 = n_2 = 2, k = 1, 2$.

Note that, by construction, each chain contains all preceding chains and that, in particular,

$$p_i^{(k-1)} = p_{i.2^{n_k}}^{(k)} .$$

We will now show that this function, defined in (6.7.4), is uniformly continuous on its domain of definition

$$\mathcal{D} = \{0_{\dot{2}}b_1b_2b_3 \dots b_s | b_i = 0 \quad \text{or} \quad 1, s = 1, 2, 3, \dots\},$$

and then extend the definition of g into $\mathcal{I}$, invoking Lemma 6.2.1 (Note that

$$0_{\dot{2}}b_1b_2b_3 \dots b_{n_1+n_2+\cdots+n_k} = \frac{i}{2^{n_1+n_2+\cdots+n_k}},$$

if $i = b_{n_1+\cdots+n_k} + 2b_{n_1+\cdots+n_k-1} + 4b_{n_1+\cdots+n_k-2} + \cdots + 2^{n_1+\cdots+n_k-1}b_1$.) Let us choose an arbitrary $\epsilon > 0$ and a k_0 so large that $\sum_{k=k_0}^{\infty} \epsilon_k < \frac{\epsilon}{2}$. We will see that $|t' - t''| < 1/2^{n_1+\cdots+n_{k_0}} \triangleq \delta, t', t'' \in \mathcal{D}$, implies that $\| g(t') - g(t'') \| < \epsilon$. Since δ is independent of t, the continuity will be uniform. Here is how we go about it: If $|t' - t''| < \delta$, then, t', t'' lie, at worst, in two adjacent intervals

$$\left[\frac{j-1}{2^{n_1+\cdots+n_{k_0}}}, \frac{j}{2^{n_1+\cdots+n_{k_0}}}\right], \quad \left[\frac{j}{2^{n_1+\cdots+n_{k_0}}}, \frac{j+1}{2^{n_1+\cdots+n_{k_0}}}\right].$$

We assume w.l.o.g. that t' lies in the first and t'' in the second interval. Let

$$g(t') = p_{j^*}^{(k_0+r)}$$

for some r, j^*. We will now trace the history of that point. It lies between the two points $p_{j_0}^{(k_0)}$ and $p_{j_0+1}^{(k_0)}$ (where $j_0 = j - 1$) which, by construction, are connected by a compact set $\mathcal{M}_{j_0+1}^{(k_0)} \subseteq N_{\epsilon_{k_0}}(p_{j_0}^{(k_0)})$. In turn, it will lie between the two points

$$p_{j_1}^{(k_0+1)}, p_{j_1+1}^{(k_0+1)} \in \mathcal{M}_{j_0+1}^{(k_0)},$$

which are connected by a compact set $\mathcal{M}_{j_1+1}^{(k_0+1)} \subseteq N_{\epsilon_{k_0+1}}(p_{j_1}^{(k_0+1)})$, and between the two points

$$p_{j_2}^{(k_0+2)}, p_{j_2+1}^{(k_0+2)} \in \mathcal{M}_{j_1+1}^{(k_0+1)},$$

which are connected by a compact set $\mathcal{M}_{j_2+1}^{(k_0+2)} \subseteq N_{\epsilon_{k_0+2}}(p_{j_2}^{(k_0+2)})$, etc.... We have $\| p_{j_1}^{(k_0+1)} - p_{j_0}^{(k_0)} \| < \epsilon_{k_0}$ and $\| p_{j_2}^{(k_0+2)} - p_{j_1}^{(k_0+1)} \| < \epsilon_{k_0+1}$ and, hence, $\| p_{j_2}^{(k_0+2)} - p_{j_0}^{(k_0)} \| < \epsilon_{k_0} + \epsilon_{k_0+1}$. Continuation of this process will eventually lead to

$$p_{j^*}^{(k_0+r)} \in N_{\epsilon_{k_0}+\cdots+\epsilon_{k_0+r-1}}(p_{j_0}^{(k_0)}),$$

i.e., $\quad \| g(t') - g(t_{j_0,k_0}) \| = \| p_{j^*}^{(k_0+r)} - p_{j_0}^{(k_0)} \| < \epsilon_{k_0} + \cdots + \epsilon_{k_0+r-1} < \epsilon/2.$

We can show, in exactly the same manner, that

$$\| g(t'') - g(t_{j_0,k_0}) \| < \epsilon/2$$

in order to obtain the desired result $\| g(t') - g(t'') \| < \epsilon$.

To extend this function to $\mathcal{I}$, we observe that $\overline{\mathcal{D}} = \mathcal{I}$ where $\overline{\mathcal{D}}$ denotes the closure of $\mathcal{D}$. Then, by Lemma 6.2.1, there is a (unique) continuous function $f : \mathcal{I} \longrightarrow M$ such that $f(t) = g(t)$ for all $t \in \mathcal{D}$ and, by (6.7.4), $f(0) = p'$, $f(1) = p''$. $\square$

We can do better than Theorem 6.7.1: If the points that are to be connected are very close together, then the connecting path remains near either point and the measure of nearness is uniform. More precisely:

(6.7.2) Theorem. *If $\mathcal{M}$ is compact, connected, and locally connected, then there is for all $\epsilon > 0$ an $\eta > 0, (\eta \leq \epsilon)$ such that, whenever $\| p' - p'' \| < \eta$, where $p', p'' \in \mathcal{M}$, there is a continuous path from p' to p'' that lies in $N_\epsilon(p') \cap \mathcal{M}$ as well as in $N_\epsilon(p'') \cap \mathcal{M}$.*

Proof. Choose $\epsilon > 0$ and let $0 < \zeta < \epsilon/2$. By Lemma 6.7, there is an $\eta > 0$ such that, whenever $\| p' - p'' \| < \eta$, there is a compact and connected set $\mathcal{M}' \subseteq \mathcal{M}$ which contains p', p'' and $\mathcal{M}' \subseteq N_\zeta(p')$ as well as $\mathcal{M}' \subseteq N_\zeta(p'')$. Let us now choose for the ϵ_k in the proof of Theorem 6.7.1

$$\epsilon_k = \zeta/2^k$$

and pick the η_1-chain from p' to p'' in $\mathcal{M}'$. We have, for the sets $\mathcal{M}_i^{(1)}$ of the proof of Theorem 6.7.1, that $\mathcal{M}_i^{(1)} \subseteq N_{\zeta/2}(p_i^{(1)})$. Since $p_i^{(1)} \in N_\zeta(p'), p_i^{(1)} \in N_\zeta(p'')$, it follows that the set $\mathcal{M}_i^{(1)}$ lies in a $(\zeta + \zeta/2)$-neighborhood of p' as well as p''. We continue this argument and eventually obtain that the sets $\mathcal{M}_i^{(k)}$ lie in a $(\zeta + \zeta/2 + \zeta/4 + \cdots + \zeta/2^k)$-neighborhood of p' as well as p''. Since the path from p' to p'' consists of points from the sets $\mathcal{M}_i^{(k)}$ and accumulation points of such points, it lies within $(\zeta + \zeta/2 + \zeta/4 + \cdots) = 2\zeta = \epsilon$ from p' as well as p'', which was to be proved. Since p', p'' are part of that path, $\eta \leq \epsilon$. □

After having proved this theorem, Hahn noted, somewhat pointedly, that he could have initially defined local connectedness by the properties that are asserted here (see Hahn [3], p. 2440) and, by implication, could have saved himself a lot of work. This is what Mazurkiewicz [1], [2] has essentially done. (See also Hausdorff [1], p. 179.)

6.8. The Hahn-Mazurkiewicz Theorem

We are now ready to construct a continuous mapping from $\mathcal{I}$ onto any compact, connected, and locally connected set $\mathcal{M} \subseteq \mathbb{E}^n$, using Hausdorff's theorem, by generalizing the extension process that led from the mapping (5.1.2) to Lebesgue's space-filling curve (Hahn [7]).

(6.8) Theorem. *(Hahn-Mazurkiewicz): A set $\mathcal{M}$ is the continuous image of $\mathcal{I}$ if and only if it is compact, connected, and locally connected.*

Proof. The necessity of the condition was established in the corollaries to Theorems 6.2.2, 6.3.1, and 6.3.2, and Theorem 6.5. We only need to prove the sufficiency.

By Hausdorff's Theorem 6.6, there is a continuous function $f : \Gamma \overset{\text{onto}}{\longrightarrow} \mathcal{M}$. Γ^c is the union of open intervals (a, b) (see also Section 5.2). We extend the definition of f to $\mathcal{I}$ by joining the images $f(a)$ and $f(b)$, subject to later modification, by a continuous path that lies in $\mathcal{M}$. (This can be done by Theorem 6.7.1.) After we have done this with every interval (a, b) in Γ^c, we wind up with a function $f : \mathcal{I} \longrightarrow \mathcal{M}$ which is surjective because its restriction to Γ is already surjective. Its restriction to Γ is continuous by Theorem 6.6 and, since Γ is compact, it is uniformly continuous on Γ. Its restriction to Γ^c is continuous by construction. That it is, in fact, continuous on $\mathcal{I}$ will now be shown by following the proof of Theorem 5.4.1 of the continuity of Lebesgue's space-filling curve.

Suppose $t_0 \in \Gamma^c$. Then, f is a continuous at t_0 by construction.

Suppose $t_0 \in \Gamma$. Then, it is a left accumulation point or a right accumulation point or a two-sided accumulation point of Γ. Let us assume that it is a left accumulation point, which makes it a right endpoint of one of the removed intervals and f is, therefore, continuous on the left. We need to show that it is also continuous on the right.

By Theorem 6.7.2, there is, for every $\epsilon > 0$, an $\eta > 0$ such that, whenever $p, p'' \in \mathcal{M}$ and $\| p' - p'' \| < \eta$, there is a continuous path from p' to p'' which lies entirely in $N_{\epsilon/2}(p') \cap \mathcal{M}$ as well as in $N_{\epsilon/2}(p'') \cap \mathcal{M}$. As we have noted in Lemma 6.7, $\eta \leq \epsilon/2$ (we are now using $\epsilon/2$ instead of ϵ).

Since $f : \Gamma \longrightarrow \mathcal{M}$ is uniformly continuous on Γ, there is a $\delta > 0$ such that

$$\| f(t_1) - f(t_2) \| < \eta \ (\leq \epsilon/2) \tag{6.8.1}$$

whenever $t_1, t_2 \in \Gamma$ and $|t_1 - t_2| < \delta$. Now, if $t \in \Gamma$, then, by (6.8.1),

$$\| f(t) - f(t_0) \| < \epsilon/2 \quad \text{for} \quad t, t_0 \in \Gamma, |t - t_0| < \delta. \tag{6.8.2}$$

If, on the other hand, $t \in \Gamma^c$, then t lies in some interval (a, b) that has been removed in the construction of the Cantor set $\Gamma : a, b \in \Gamma, (a, b) \subset \Gamma^c$. If $b - a < \delta$, then, by (6.8.1), $\| f(a) - f(b) \| < \eta$ and there exists, by Theorem 6.7.2, a continuous path from $f(a)$ to $f(b)$ that lies in an $\epsilon/2$-neighborhood of $f(a)$. If we denote the path by $f : (a, b) \longrightarrow \mathcal{M}$, then $\| f(t) - f(a) \| < \epsilon/2$ for all $t \in (a, b)$. Hence, if we also observe (6.8.1) with t_0 for t_1 and a for t_2, we have

$$\begin{aligned} \| f(t) - f(t_0) \| &\leq \| f(t) - f(a) \| + \| f(a) - f(t_0) \| \\ &< \epsilon/2 + \epsilon/2 = \epsilon. \end{aligned} \tag{6.8.3}$$

The proof for a right accumulation point is analogous and the proof for a two-sided accumulation point can be put together from these two proofs. □

This proof of the Hahn-Mazurkiewicz theorem may be viewed as a recipe for the construction of space-filling curves of the Lebesgue type. Lebesgue's space-filling curve itself is obtained by taking $\mathcal{Q}$ for the target set $\mathcal{M}$ and using the subsquares $\mathcal{Q}_{00}, \mathcal{Q}_{01}, \mathcal{Q}_{10}, \mathcal{Q}_{11}$ of Fig. 5.4.1 for the sets $\mathcal{K}_{i_1}$ ($i_1 = 0, 1, 2, 3$), the subsquares $\mathcal{Q}_{0000}, \mathcal{Q}_{0001}, \mathcal{Q}_{0010}, \mathcal{Q}_{0011}$, etc.... of Fig. 5.5.2 for the sets $\mathcal{K}_{i_1 i_2}$ ($i_2 = 0, 1, 2, 3$), etc.... and then joining the exit point from each square to the entry point of the following square by a straight line segment. Hahn's first proof (in Hahn [3]) is, by contrast, a generalization of Peano's construction.

Tracing the way through the definitions, lemmas, and theorems that ultimately led to the Hahn-Mazurkiewicz theorem, one comes to the realization that the only relevant concepts are *open sets (neighborhoods), continuity, closeness* (which requires a metric of sorts), and *connectedness.* So, in a topological space (a space $\mathcal{S}$ which contains a collection τ of subsets—open sets—with the property that any union of elements of τ is again in τ, any finite intersection of elements of τ is in τ, and $\emptyset$ and $\mathcal{S}$ are in τ) that is *Hausdorff* (with any two distinct points $p_1, p_2 \in \mathcal{S}$ there are disjoint open sets $\Omega_1, \Omega_2 \in \tau$ such that $p_1 \in \Omega_1, p_2 \in \Omega_2$), the theorem ought to be valid. Indeed, one can show that *a Hausdorff space is a continuous image of $\mathcal{I}$ if and only if it is a compact, connected, and locally connected metric space* (Willard [1], pp. 23, 86, 219, 221). Hahn himself demonstrated (in Hahn [3], pp. 2475–2489) that his result was still valid in *Fréchet-spaces.* These are abstract spaces with a topological structure, which Fréchet introduced in his 1906 thesis (Fréchet [1]). They were harbingers of Hausdorff spaces but receded into the background when the latter emerged in 1914 (Engelking [1]).

In addition to Schoenflies' characterization, which we mentioned at the beginning of this chapter, there appeared other characterizations as well. In 1916, Mazurkiewicz characterized the continuous image of $\mathcal{I}$ as compact connected sets $\mathcal{M}$ all points of which are *of the first kind*, meaning that for any $p \in \mathcal{M}$ and all $\epsilon > 0$, there is a compact connected subset $\mathcal{K}$ of $\mathcal{M}$ with diameter less than ϵ, and an $\eta > 0$ such that whenever a point from $\mathcal{M}$ lies within η of p it lies in $\mathcal{K}$. (See Mazurkiewicz [5], p. 441 or Sierpiński [8], pp. 56–57.) In 1921, Hahn arrived at the same characterization (Hahn [6]). The same year, Hahn demonstrated (Hahn [5]) that for every component of an open set in a metric space S to be open, it is necessary and sufficient that S be locally connected. (For $p \in S$, the union of all connected subsets of S that contain p is called a component of S.)

In 1920, Sierpiński found still another characterization: *A set is a continuous image of $\mathcal{I}$ if and only if it is compact, connected, and can be represented as the union of a finite number of compact and connected sets with diameters $\leq \delta$ for every $\delta > 0$* (Sierpiński [8]). (The reader is advised that when Sierpinski and Mazurkiewicz use the term *Jordan Curve*, they mean a continuous image of a line segment and not a continuous injective image of a line segment.)

In 1921, A. Rosenthal constructed an example of a bounded simply connected set that is the continuous image of a line segment and whose boundary is not a continuous curve (Rosenthal [1]).

6.9. Generation of Space-Filling Curves by Stochastically Independent Functions

We have encountered two methods of generating space-filling curves: On the one hand, we had the method of partitioning the target set into b^{2n} replicas of itself and establishing a correspondence between the b^{2n} congruent subintervals of $\mathcal{I}$ and the aforementioned replicas, arranged in a suitable order, which led to Hilbert-type curves such as the Hilbert curve, the Peano curve, the Sierpiński curve, and variations thereof. On the other hand, we had the method that is outlined in Hahn's proof of Theorem 6.8, which leads to n-dimensional (and even infinite dimensional) space-filling curves of which the Lebesgue curve (of Chapter 5) and the Schoenberg-curve (which will be discussed in the next chapter) are two-dimensional examples.

In 1936, Hugo Steinhaus made the surprising discovery that space-filling curves may be generated by stochastically independent functions (Steinhaus [1]). He published his result in a tersely written note, buried deeply within the 2331-page Volume 202 of the *Comptes Rendus de l'Academie des Sciences, Paris* (1936), where it was largely ignored. Forty years later, A.M. Garsia and others, unaware of what Steinhaus had done, arrived at the same conclusion, elaborated upon it, embroidered it, and disseminated it in a series of publications (Garsia [1], Gelbaum [1], Bosznay [1], Milne [1], Bryc [1], Holbrook [1], Yost [1], Robertson [1]—see also Donoghue [1]).

The significance of Steinhaus' result goes well beyond what meets the eye because it allows for the construction of n-dimensional (and $\aleph_0$-dimensional) space-filling curves in terms of the known coordinate functions of a two-dimensional space-filling curve such as the Hilbert curve and the Peano curve.

(6.9.1) Definition. *n measurable functions $\varphi_1, \varphi_2, \ldots, \varphi_n : \mathcal{I} \longrightarrow \mathbb{R}$ are called stochastically independent with respect to the Lebesgue measure if, for any n measurable sets $\mathcal{A}_1, \mathcal{A}_2, \ldots, \mathcal{A}_n \subset \mathbb{R}$,*

$$\Lambda_1\left[\bigcap_{j=1}^{n} \varphi_j^*(\mathcal{A}_j)\right] = \prod_{j=1}^{n} \Lambda_1[\varphi_j^*(\mathcal{A}_j)]. \tag{6.9.1}$$

If such functions are also continuous and non-constant, they define an n-dimensional space-filling curve:

(6.9.1) Theorem. *(Steinhaus): If $\varphi_1, \varphi_2, \ldots, \varphi_n : \mathcal{I} \longrightarrow \mathbb{R}$ are continuous, non-constant, and stochastically independent with respect to the Lebesgue measure, then*

$$f = (\varphi_1, \varphi_2, \ldots, \varphi_n)^T : \mathcal{I} \overset{onto}{\longrightarrow} \varphi_{1*}(\mathcal{I}) \times \varphi_{2*}(\mathcal{I}) \times \cdots \times \varphi_{n*}(\mathcal{I})$$

is a space-filling curve.

Proof. Since $\varphi_1, \varphi_2, \ldots, \varphi_n$ are continuous, we have, from the Corollary to Theorems 6.3.1 and 6.3.2, that $\varphi_{1*}(\mathcal{I}), \varphi_{2*}(\mathcal{I}), \ldots \varphi_{n*}(\mathcal{I})$ are closed intervals, and since $\varphi_1, \varphi_2, \ldots \varphi_n$ are non-constant, these intervals have interior points. Let $\mathcal{P} \triangleq \varphi_{1*}(\mathcal{I}) \times \varphi_{2*}(\mathcal{I}) \times \cdots \times \varphi_{n*}(\mathcal{I})$ and let $x = (\xi_1, \xi_2, \ldots, \xi_n)^T \in \mathcal{P}$. $\mathcal{P}$, the cartesian product of compact sets, is compact. Let $\mathcal{A}_j = (\xi_j - \epsilon, \xi_j + \epsilon)$, $j = 1, 2, 3, \ldots, n$ for an arbitrary $\epsilon > 0$. By construction, $\varphi_j^*(\mathcal{A}_j) \neq \emptyset$ for all $j = 1, 2, 3, \ldots, n$. Since $\varphi_1, \varphi_2, \ldots, \varphi_n$ are continuous, there are, by Theorem 6.1.1, open sets $\mathcal{A}_j^{(1)} \subseteq \mathbb{R}$ such that $\varphi_j^*(\mathcal{A}_j) = \mathcal{A}_j^{(1)} \cap \mathcal{I}$ and, hence, all these sets are measurable and $\Lambda_1[\varphi_j^*(\mathcal{A}_j)] > 0$. Hence, by (6.9.1), $\Lambda_1[\bigcap_{j=1}^n \varphi_j^*(\mathcal{A}_j)] > 0$ and, consequently, there is some $t \in \bigcap_{j=1}^{n} \varphi_j^*(\mathcal{A}_j)$. This means, in turn, that $f(t) \in \mathcal{A}_1 \times \mathcal{A}_2 \times \cdots \times \mathcal{A}_n$, i.e., $\| f(t) - x \| < \sqrt{n}\epsilon$.

Therefore, $f_*(\mathcal{I})$ is dense in $\mathcal{P}$. By the Corollary to Theorem 6.2.2, $f_*(\mathcal{I})$ is also compact. Since $\mathcal{P}$ is compact, we have from Lemma 6.2.2 that $f_*(\mathcal{I}) = \mathcal{P}$. □

In order to use this result for the construction of higher-dimensional space-filling curves in terms of known two-dimensional space-filling curves, we also need the concept of *uniform distribution*:

(6.9.2) Definition. *A function $\varphi : \mathcal{I} \longrightarrow \mathbb{R}$ is said to be uniformly distributed with respect to the Lebesgue measure if for any measurable set $\mathcal{A} \subset \mathbb{R}$,*

$$\Lambda_1[\varphi^*(\mathcal{A})] = \Lambda_1(\mathcal{A}). \tag{6.9.2}$$

Steinhaus suggested the following construction of a four-dimensional space-filling curve: Suppose that $\varphi, \psi : \mathcal{I} \longrightarrow \mathcal{I}$ are continuous, non-constant, stochastically independent, and uniformly distributed.

Let

$$\alpha(t) = \varphi(\varphi(t)), \beta(t) = \varphi(\psi(t)), \gamma(t) = \psi(\varphi(t)), \delta(t) = \psi(\psi(t)). \tag{6.9.3}$$

The functions $\alpha, \beta, \gamma, \delta$ are continuous, non-constant, and uniformly distributed. They are also stochastically independent, as we will see from the following argument:

Let $\mathcal{A}, \mathcal{B}, \mathcal{C}, \mathcal{D}$ denote measurable subsets of $\mathbb{R}$ and let

$$\mathcal{A}_1 = \varphi^*(\mathcal{A}), \mathcal{B}_1 = \varphi^*(\mathcal{B}), \mathcal{C}_1 = \psi^*(\mathcal{C}), \mathcal{D}_1 = \psi^*(\mathcal{D})$$
$$\mathcal{A}_2 = \varphi^*(\mathcal{A}_1), \mathcal{B}_2 = \psi^*(\mathcal{B}_1), \mathcal{C}_2 = \varphi^*(\mathcal{C}_1), \mathcal{D}_2 = \psi^*(\mathcal{D}_1).$$

From (6.1.5),

$$\begin{aligned} \alpha^*(\mathcal{A}) &= \varphi^*(\varphi^*(\mathcal{A})), \beta^*(\mathcal{B}) = \psi^*(\varphi^*(\mathcal{B})), \\ \gamma^*(\mathcal{C}) &= \varphi^*(\psi^*(\mathcal{C})), \delta^*(\mathcal{D}) = \psi^*(\psi^*(\mathcal{D})). \end{aligned} \tag{6.9.4}$$

We have, because of the stochastic independence and uniform distribution of φ, ψ, and in view of (6.1.3) and (6.9.4), that

$$\begin{aligned} \Lambda_1[\alpha^*(\mathcal{A}) \cap \beta^*(\mathcal{B}) \cap \gamma^*(\mathcal{C}) \cap \delta^*(\mathcal{D})] &= \Lambda_1(\mathcal{A}_2 \cap \mathcal{B}_2 \cap \mathcal{C}_2 \cap \mathcal{D}_2) \\ &= \Lambda_1[\varphi^*(\mathcal{A}_1) \cap \psi^*(\mathcal{B}_1) \cap \varphi^*(\mathcal{C}_1) \cap \psi^*(\mathcal{D}_1)] \\ &= \Lambda_1[\varphi^*(\mathcal{A}_1 \cap \mathcal{C}_1) \cap \psi^*(\mathcal{B}_1 \cap \mathcal{D}_1)] \\ &= \Lambda_1[\varphi^*(\mathcal{A}_1 \cap \mathcal{C}_1)]\Lambda_1[\psi^*(\mathcal{B}_1 \cap \mathcal{D}_1)] \\ &= \Lambda_1(\mathcal{A}_1 \cap \mathcal{C}_1)\Lambda_1(\mathcal{B}_1 \cap \mathcal{D}_1) \\ &= \Lambda_1[\varphi^*(\mathcal{A}) \cap \psi^*(\mathcal{C})]\Lambda_1[\varphi^*(\mathcal{B}) \cap \psi^*(\mathcal{D})] \\ &= \Lambda_1[\varphi^*(\mathcal{A})]\Lambda_1[\psi^*(\mathcal{C})]\Lambda_1[\varphi^*(\mathcal{B})]\Lambda_1[\psi^*(\mathcal{D})] \\ &= \Lambda_1(\mathcal{A}_1)\Lambda_1(\mathcal{B}_1)\Lambda_1(\mathcal{C}_1)\Lambda_1(\mathcal{D}_1) \\ &= \Lambda_1[\varphi^*(\mathcal{A}_1)]\Lambda_1[\psi^*(\mathcal{B}_1)]\Lambda_1[\varphi^*(\mathcal{C}_1)]\Lambda_1[\psi^*(\mathcal{D}_1)] \\ &= \Lambda_1(\mathcal{A}_2)\Lambda_1(\mathcal{B}_2)\Lambda_1(\mathcal{C}_2)\Lambda_1(\mathcal{D}_2) \\ &= \Lambda_1[\alpha^*(\mathcal{A})]\Lambda_1[\beta^*(\mathcal{B})]\Lambda_1[\gamma^*(\mathcal{C})]\Lambda_1[\delta^*(\mathcal{D})], \end{aligned}$$

i.e., $\alpha, \beta, \gamma, \delta$ are stochastically independent and define, by Theorem 6.9 a four-dimensional space-filling curve.

Such functions φ, ψ, as needed for the preceding construction may be obtained as the coordinate functions of measure-preserving maps $f : \mathcal{I} \overset{\text{onto}}{\longrightarrow} \mathcal{Q}$:

(6.9.3) Definition. $f : \mathcal{I} \overset{onto}{\longrightarrow} \mathcal{Q}$ *is called measure-preserving if, for any measurable subset* $\mathcal{A} \subseteq \mathcal{Q}$,

$$\Lambda_1(f^*(\mathcal{A})) = \Lambda_2(\mathcal{A}).$$

We have

(6.9.2) Theorem. *If* $f = \begin{pmatrix} \varphi \\ \psi \end{pmatrix} : \mathcal{I} \overset{onto}{\longrightarrow} \mathcal{Q}$ *is measure-preserving, then its coordinate functions* φ, ψ *are uniformly distributed and stochastically independent.*

Proof. (We owe the following simple proof to Krzysztof Loskot of the Polskiej Akademii Nauk in Katowice.)

From $f^*(\mathcal{A} \times \mathcal{B}) = \{t \in \mathcal{I} | \varphi(t) \in \mathcal{A} \;\&\; \psi(t) \in \mathcal{B}\} = \varphi^*(\mathcal{A}) \cap \psi^*(\mathcal{B})$ for any subsets $\mathcal{A}, \mathcal{B} \subseteq \mathcal{I}$, we obtain for $\mathcal{B} = \mathcal{I}$,

$$f^*(\mathcal{A} \times \mathcal{I}) = \varphi^*(\mathcal{A}) \cap \psi^*(\mathcal{I}) = \varphi^*(\mathcal{A})$$

and, hence, since f is measure-preserving,

$$\begin{aligned} \Lambda_1(\varphi^*(\mathcal{A})) &= \Lambda_1(f^*(\mathcal{A} \times \mathcal{I})) = \Lambda_2(\mathcal{A} \times \mathcal{I}) \\ &= \Lambda_1(\mathcal{A})\Lambda_1(\mathcal{I}) = \Lambda_1(\mathcal{A}). \end{aligned}$$

By analogous reasoning,

$$\Lambda_1(\psi^*(\mathcal{B})) = \Lambda_1(\mathcal{B}),$$

i.e., φ, ψ are uniformly distributed. Since

$$\begin{aligned} \Lambda_1(\varphi^*(\mathcal{A}) \cap \psi^*(\mathcal{B})) &= \Lambda_1(f^*(\mathcal{A} \times \mathcal{B})) = \Lambda_2(\mathcal{A} \times \mathcal{B}) = \Lambda_1(\mathcal{A})\Lambda_1(\mathcal{B}) \\ &= \Lambda_1(\varphi^*(\mathcal{A}))\Lambda_1(\psi^*(\mathcal{B})), \end{aligned}$$

we see that φ, ψ are also stochastically independent. □

(The converse of this theorem is also true but it is not needed for our purposes.)

The Hilbert- , Peano- , and Sierpiński-Knopp maps of Chapters 2, 3, and 4 are, in fact, measure-preserving. In his 1976 paper, Garsia notes that "this fact seems to have gone unnoticed in the literature so far" (Garsia [1], p. 167, second paragraph). It was, in fact, known to Steinhaus 40 years earlier and, quite possibly, before then. For a proof that the Peano-map is measure-preserving, we refer the reader to a 1980 paper by S.C. Milne (Milne [1], pp. 155–157).

With φ, ψ stochastically independent and uniformly distributed, such as, for example, the coordinate functions of the Peano curve,

$$\left.\begin{array}{l} x_1 = \varphi(t) \\ x_2 = \varphi \circ \psi(t) \\ x_3 = \varphi \circ \psi \circ \psi(t) \\ \vdots \\ x_n = \varphi \circ \psi \circ \psi \circ \ldots \circ \psi(t) \end{array}\right\} \quad t \in \mathcal{I}$$

defines an n-dimensional space-filling curve, as the reader can easily confirm, because the coordinate functions are continuous, non-constant, and stochastically independent. The idea extends, quite naturally, to $\aleph_0$-dimensional space-filling curves, as Steinhaus noted in his paper. (See also Jessen [1].) The implication of this is that the cardinality of the $\aleph_0$-dimensional unit cube $\mathcal{I} \times \mathcal{I} \times \mathcal{I} \times \cdots$ is $\mathfrak{c}$, the cardinality of the continuum. This was already known to Lebesgue, who, in a 1905 paper (Lebesgue [2], p. 210), showed that there exists a sequence of continuous functions $\varphi_n : \mathcal{I} \to \mathcal{I}, n = 1, 2, 3, \ldots$, such that, for any given sequence of real numbers $a_n \in \mathcal{I}$, there is at least one $t \in \mathcal{I}$ such that $\varphi_n(t) = a_n, n = 1, 2, 3, \ldots$. In a 1937 paper, Sierpiński has shown that functions φ_n, satisfying Lebesgue's theorem, may be represented in terms of the coordinate functions φ, ψ of the Peano curve by $\varphi_n(t) = \varphi \circ \psi \circ \ldots \circ \psi(t)$ where the composition of ψ is to be taken $n - 1$ times—the same result that Steinhaus achieved a year earlier by a different route (Sierpiński [9]; See also Kuratowski [1], p. 150, and Dinghas [1].) In Section 7.5, we will encounter a different construction of an $\aleph_0$—dimensional space-filling curve.

6.10. Representation of a Space-Filling Curve by an Analytic Function

In 1945, R. Salem and A. Zygmund found complex-valued functions of a complex variable, analytic on the open unit disk and continuous on the closed unit disk, which map the unit disk's circumference onto a square (Salem-Zygmund [1]). Their functions are represented by lacunary series. There was no response until seven years later in 1952, when G. Piranian, C.T. Titus, and G.S. Young published an ingenious, yet quite elementary construction of such a function (Piranian, Titus, and Young [1]), to be followed by the results of A.C. Schaeffer in 1954 (Schaeffer [1]) and by F. Bagemihl and G. Piranian in 1960 (Bagemihl and Piranian [1]).

In the following theorem, we present the 1952 result of G. Piranian, C.J. Titus, and G.S. Young:

(7.10) Theorem. *There is a complex-valued function of a complex variable that is analytic on the open unit disk $|z| < 1$ and continuous on the closed unit disk $|z| \leq 1$, which maps the circumference $|z| = 1$ onto the square $[-1, 1]^2$.*

Proof. (by construction): We propose to demonstrate that the function

$$f(z) = \sum_{k=1}^{\infty} \gamma_k[1 - (1 - z/\beta_k)^{\alpha_k}], \alpha_k > 0, \beta_k, \gamma_k \in \mathbb{C} \tag{6.10.1}$$

has all the required properties, provided the sequences $\{\alpha_k\} \longrightarrow 0, \{\gamma_k\} \rightarrow 0$, and the sequence of distinct points $\{\beta_k\}$ on the unit circle $|z| = 1$ are suitably chosen.

Let

$$f_k(z) = \gamma_k[1 - (1 - z/\beta_k)^{\alpha_k}]. \tag{6.10.1}$$

With $|\beta_k| = 1, |z| \leq 1$, and $\alpha_k > 0$, we have in view of $|1 - z/\beta_k| = (1/|\beta_k|)|\beta_k - z| \leq |\beta_k| + |z| \leq 2$ that

$$\begin{aligned} |f_k(z)| &= |\gamma_k| \, |1 - (1 - z/\beta_k)^{\alpha_k}| \\ &\leq |\gamma_k|(1 + e^{\alpha_k \log|\beta_k - z|}) \leq |\gamma_k|(1 + e^{\alpha_k \log 2}) \leq 4|\gamma_k| \end{aligned} \tag{6.10.2}$$

for all $|z| \leq 1$ for sufficiently small α_k.

Let $\mathcal{N}_k \triangleq N_{\delta_k}(\beta_k)$ and let $|z| \leq 1$ but $z \in \mathcal{N}_k^c$. Since

$$(1 - z/\beta_k)^{\alpha_k} = e^{\alpha_k (1/\beta_k) \log(\beta_k - z)},$$

and since $|\beta_k - z| \geq \delta_k$, we can choose α_k so small that $(1 - z/\beta_k)^{\alpha_k}$ is as close to 1 as we please, and hence: For suitable small α_k,

$$|f_k(z)| \leq 1/k^2 \quad \text{for all} \quad z \in \mathcal{N}_k^c. \tag{6.10.3}$$

(Note that $\{\gamma_k\} \longrightarrow 0$ and, hence, $|\gamma_k| \leq M$ for some $M > 0$ and all k.)

We are now ready for the choice of α_k, β_k, γ_k: We partition the square $[-1, 1]^2$ into four congruent subsquares $\mathcal{Q}_1, \mathcal{Q}_2, \mathcal{Q}_3, \mathcal{Q}_4$, as indicated in Fig. 6.10, and let $\beta_1 = (\sqrt{2}/2)(1 - i)$, $\beta_2 = (\sqrt{2}/2)(-1 + i)$, $\beta_3 = (\sqrt{2}/2)(1 + i)$, and $\beta_4 = (\sqrt{2}/2)(1 - i)$, $\gamma_k = \beta_k$ for $k = 1, 2, 3, 4$. Then, $f_k(\beta_k) = \gamma_k(= \beta_k)$ for $k = 1, 2, 3, 4$. Pick $\mathcal{N}_k$ so that $\mathcal{N}_k$ does not contain $\beta_1, \beta_2, \ldots \beta_{k-1}$ and choose $\alpha_1, \alpha_2, \alpha_3, \alpha_4$ to be so small that

$$F_4(\beta_k) = f_1(\beta_k) + f_2(\beta_k) + f_3(\beta_k) + f_4(\beta_k) \in \mathcal{Q}_k^0,$$

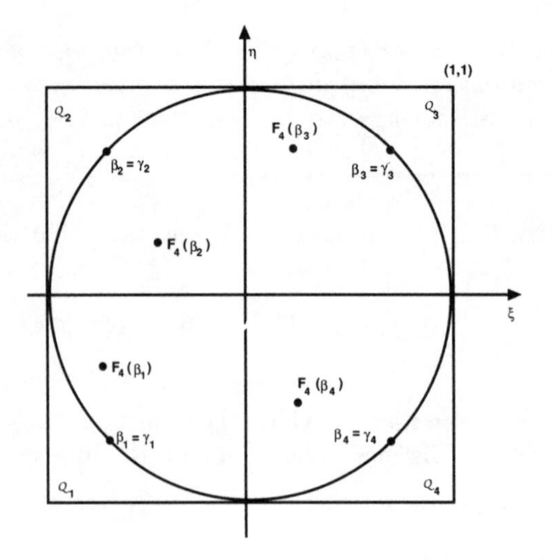

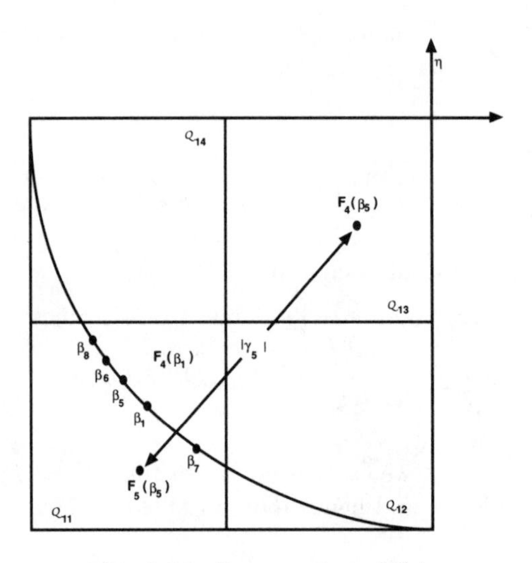

Fig. 6.10. Construction of f(z)

and (6.10.3) is satisfied for $k = 1, 2, 3, 4$. Since z either lies in one of the $\mathcal{N}_k$ or in none of them, we have, in view of (6.10.2) and (6.10.3),

$$\begin{aligned} |f_1(z)| + f_2(z)| + |f_3(z)| + |f_4(z)| &\leq 4|\gamma_j| + 1 + 1/2^2 + 1/3^2 + 1/4^2 \\ &= 4 + \sum_{k=1}^{4} 1/k^2 \end{aligned}$$

for some $j \in \{1, 2, 3, 4\}$. (Note that $|\gamma_j| = 1$ for $j = 1, 2, 3, 4$.)

Next, we partition the squares $\mathcal{Q}_k$ into four congruent subsquares $\mathcal{Q}_{jk}$, and arrange them in the same order as when constructing the Hilbert curve in Fig. 2.1. (see Fig. 6.10) and pick $\beta_5, \beta_6, \beta_7, \beta_8$ (near β_1) so that $F_4(\beta_k) \in \mathcal{Q}_1$ for $k = 5, 6, 7, 8$. We choose γ_5 so that $F_5(\beta_5) \in \mathcal{Q}_{11}$. This is always possible because $F_5 = F_4 + f_5$ and $F_4(\beta_5) \in Q_1$. Therefore, all we have to do is pick $|\gamma_4| = |f_5(\beta_5)| = |F_5(\beta_5) - F_4(\beta_5)| \leq \sqrt{2}$ which is the length of the diagonal of Q_1 (see Fig. 6.10). Finally, choose α_5 that is so small that $F_6(e^{i\theta}), 0 \leq \theta \leq 2\pi$, still passes through the interiors of $\mathcal{Q}_2, \mathcal{Q}_3, \mathcal{Q}_4$ and (6.10.3) is satisfied for $k = 6$. We proceed in this manner to obtain $F_7(\beta_7) \in \mathcal{Q}_{21}, F_8(\beta_8) \in \mathcal{Q}_{22}$ with $F_8(e^{i\theta})$ still passing though the interiors of $\mathcal{Q}_2, \mathcal{Q}_3, \mathcal{Q}_4$ and (6.10.3) satisfied for $k = 7, 8$. Next, we partition $\mathcal{Q}_2$, pick $\beta_9, \beta_{10}, \beta_{11}, \beta_{12}$ (near β_2), and proceed as before. After having partitioned each of the four squares $\mathcal{Q}_1, \mathcal{Q}_2, \mathcal{Q}_3, \mathcal{Q}_4$ into four subsquares and picked β_5 to β_{20}, γ_5 to γ_{20}, and α_5 to α_{20} so that $F_{20}(e^{i\theta})$ passes through the interiors of all the subsquares $\mathcal{Q}_{jk}$ and (6.10.2) is satisfied where each $\mathcal{N}_k$ is chosen so that it does not contain $\beta_1, \beta_2, \ldots, \beta_{k-1}$, we obtain

$$|f_5(z)| + |f_6(z)| + \cdots + |f_{20}(z)| \leq 4\sqrt{2} + \sum_{k=5}^{20} 1/k^2$$

because $z \in \mathcal{N}_k$ for some $k \in \{5, 6, \ldots, 20\}$ or it is not in any of them. As the next step, we partition each of the 16 subsquares Q_{jk} into four congruent subsquares each, and note that $|\gamma_k| \leq \sqrt{2}/2$, the length of the diagonal of Q_{jk}, for $k = 21, 22, \ldots, 84$, and we obtain, as before,

$$|f_{21}(z)| + |f_{22}(z)| + \cdots + |f_{84}(z)| \leq 4\sqrt{2}/2 + \sum_{k=21}^{84} 1/k^2.$$

Eventually,

$$\sum_{k=1}^{\infty} |f_k(z)| \leq 4 + 4\sqrt{2} \sum_{k=0}^{\infty} 1/2^k + \sum_{k=1}^{\infty} 1/k^2.$$

Since f_k is analytic in $|z| < 1$ and continuous in $|z| \leq 1$, we see that f, as defined in (6.10.1), is indeed analytic in $|z| < 1$, continuous in $|z| \leq 1$ (Carathéodory [1]), and, by construction, $f(e^{i\theta}), 0 \leq \theta \leq 2\pi$, passes through the interior of every subsquare, i.e., is dense in $[-1, 1]^2$. Since $\{f(e^{i\theta}) | 0 \leq \theta \leq 2\pi\}$ is also compact, we have from Lemma 6.2.2 that it is equal to $[-1, 1]^2$, and our theorem is proved. □

6.11. Problems

1. Show that the set of rational numbers is dense in $\mathbb{R}$.
2. Let $\mathcal{D} = \{0_{\dot{2}} b_1 b_2 b_3, , , b_s | b_j = 0 \text{ or } 1, s = 1, 2, 3, \ldots\}$. Show that $\mathcal{D}$ is dense in $\mathcal{I}$.

3. Show: If $\overline{\mathcal{D}} = \mathcal{I}$, then, for every $a \in \mathcal{I}$, there is a sequence $\{a_n\}, a_n \in \mathcal{D}$, such that $\lim_{n \longrightarrow \infty} a_n = a$.
4. Verify the relations (6.1.1) to (6.1.4).
5. Prove: If $\overline{\mathcal{D}} = \mathcal{I}$ and $f(t) = g(t)$ for all $t \in \mathcal{D}$, and if f, g are continuous, then $f(t) = g(t)$ for all $t \in \mathcal{I}$.
6. Find an open cover of the interval $(0, 1]$ which does not contain a finite subcover.
7. Prove: Every infinite subset of a compact set $\mathcal{K}$ contains a sequence that converges to a point in $\mathcal{K}$.
8. Prove: The cartesian product of finitely many compact sets is compact.
9. Show that the set in Fig. 6.3.1 is connected but the set in Fig. 6.3.2 is not.
10. Show: If $f : \mathcal{I} \longrightarrow \mathbb{R}$ is continuous, then $f_*(\mathcal{I})$ is an interval.
11. Prove: If two connected sets have a point in common, then their union is connected.
12. Prove Lemma 6.3 without assuming that $\mathcal{C}$ is closed.
13. Show that the set in Fig. 6.3.1 is not locally connected, but the set in Fig. 6.3.2 is.
14. Show: The set in Fig. 6.3.1 is a perfect set. (A set is *perfect* if it is equal to the set of its accumulation points.)
15. Show: The function from the set in Fig. 6.3.2 to the set in Fig. 6.3.1, which is described on p. 98, is continuous.
16. Let $\mathcal{S}_{b_1}, \mathcal{S}_{b_1,b_2}, \mathcal{S}_{b_1 b_2 b_3} \ldots$ denote non-empty compact subsets of $\mathbb{E}^n$, whereby $b_j = 0$ or 1, such that

$$\mathcal{S}_{b_1} \supseteq \mathcal{S}_{b_1 b_2} \supseteq \mathcal{S}_{b_1 b_2 b_3} \supseteq \ldots$$

with the diameter of $\mathcal{S}_{b_1 b_2 b_3 \ldots b_k}$ shrinking to 0 as $k \longrightarrow \infty$. By Cantor's intersection theorem,

$$\mathcal{S}_{b_1} \cap \mathcal{S}_{b_1 b_2} \cap \mathcal{S}_{b_1 b_2 b_3} \cap \ldots$$

consists of a single point in $\mathbb{E}^n$. The set of all such points,

$$\bigcup \left(\bigcap \mathcal{S}_{b_1 b_2 b_3 \ldots b_k} \right),$$

where the union is to be extended over all dyadic sequences, is called a *dyadic* set. Prove: Every compact set is a dyadic set.
17. If, in Problem 16, $\mathcal{S}_0, \mathcal{S}_1$ are disjoint, $\mathcal{S}_{b_1 0}, \mathcal{S}_{b_1 1}$ are disjoint, $\mathcal{S}_{b_1 b_2 0}$, $\mathcal{S}_{b_1 b_2 1}$ are disjoint, etc...., then the dyadic set of Problem 16 is called a *dyadic discontinuum*. Show: The Cantor set is a dyadic discontinuum.
18. Prove: Every compact set is a continuous image of a dyadic discontinuum.
19. Show that the sets in Fig. 6.3.1 and 6.3.2 are not pathwise connected.
20. Use the method outlined in the proof of Theorem 6.8 to construct space-filling curves $f : \mathcal{I} \overset{\text{onto}}{\longrightarrow} \mathcal{Q}$ other than Lebesgue's.

21. An *arc* is the image of a continuous, differentiable function from a closed and bounded interval into $\mathbb{E}^n$. A subset of $\mathbb{E}^n$ is *arcwise connected* if any two of its points can be connected by an arc that lies entirely in the set. A subset $\mathcal{M}$ of $\mathbb{E}^n$ is *uniformly arcwise connected* if it is arcwise connected and if, for any $\epsilon > 0$, there is an $\eta > 0$ such that for any two points $p', p'' \in \mathcal{M}$ with $\| p' - p'' \| < \eta$, there is an arc from p' to p'' that lies entirely in $N_\epsilon(p') \cap \mathcal{M}$, as well as in $N_\epsilon(p'') \cap \mathcal{M}$. Show: If the compact subset $\mathcal{M}$ of $\mathbb{E}^n$ is uniformly arcwise connected, then there is a continuous map from $\mathcal{I}$ onto $\mathcal{M}$ that is differentiable a.e.
22. Show that the functions $\alpha, \beta, \gamma, \delta$, which are defined in (6.9.3), are uniformly distributed. (See (6.9.2).)
23. Let $\varphi, \psi : \mathcal{I} \longrightarrow \mathbb{R}$ denote stochastically independent and uniformly distributed continuous non-constant functions. Show that the functions $\alpha = \varphi, \beta = \varphi \circ \psi, \gamma = \varphi \circ \psi \circ \psi$ are stochastically independent. Let $\varphi = \varphi_p, \psi = \psi_p$, and give a geometric interpretation of the mapping $f = (\alpha, \beta, \gamma)^T : \mathcal{I} \longrightarrow \mathbb{R}^3$.
24. With φ, ψ as in Problem 23, let $\alpha_1 = \varphi, \alpha_2 = \varphi \circ \psi, \alpha_3 = \varphi \circ \psi \circ \psi, \ldots, \alpha_n = \varphi \circ \psi \circ \psi \circ \ldots \circ \psi$. Show that $a = (\alpha_1, \alpha_2, \ldots, \alpha_n)^T : \mathcal{I} \longrightarrow \mathbb{R}^n$ represents an n-dimensional space-filling curve.
25. Use the result of Problem 24 to construct a three-dimensional Hilbert curve and a three-dimensional Peano curve. Compare your results with the curves in Sections 2.8 and 3.8.
26. Show: If $f : \mathcal{I} \longrightarrow \mathcal{Q}$ is measure-preserving by definition (6.9.3), then $\Lambda_2(f(\mathcal{B})) = \Lambda_1(\mathcal{B}), \mathcal{B} \subseteq \mathcal{I}$.
27. Prove the converse of Theorem 6.9.2.
28. Is the space-filling curve $f \circ e^{i\theta} : [0, 2\pi] \longrightarrow [-1, 1]^2$, where f is defined in (6.10.1), after scale-change and appropriate translation, the Hilbert curve of Section 2.1?
29. Let $\mathcal{A}_1 \supseteq \mathcal{A}_2 \supseteq \mathcal{A}_3 \supseteq \mathcal{A}_4 \supseteq \ldots$ represent compact, connected, and locally connected sets. Show by means of an example that $\bigcap_{n=1}^{\infty} \mathcal{A}_n$ need not be locally connected.
30. The *Cantor brush* is defined as the set of all points in $\mathbb{E}^2$ that lie on the line segments from the point $(1/2, 1)$ to all points of the Cantor set. Is the Cantor set compact, connected, or locally connected?
31. Show that the points in $\mathcal{Q}$ which, under Peano's mapping $f_p : [0, 1] \rightarrow \mathcal{Q}$, have four preimages in [0,1] are dense in $\mathcal{Q}$.
32. Show that the points in $\mathcal{Q}$ which, under Hilbert's mapping $f_h : [0, 1] \rightarrow \mathcal{Q}$, have at least three preimages in $[0, 1]$ are dense in $\mathcal{Q}$.
33. Do the same as in problem 32 for the Sierpiński-Knopp mapping $f_s : [0, 1] \rightarrow \mathcal{Q}$.

Chapter 7

Schoenberg's Space-Filling Curve

7.1. Definition and Basic Properties

Isaac J. Schoenberg (1903–1990) was born in Galatz, Romania, and died in Madison, Wisconsin. He studied at the Universities of Jassy, Göttingen, and Berlin, and received his doctor's degree from the University of Jassy in 1926. After emigrating to the United States in 1930, he held fellowships at the University of Chicago, Harvard, and the Institute for Advanced Studies at Princeton. From 1935 on, he held positions on the faculties of Swarthmore College, Colby College, and The University of Pennsylvania, where he became Professor in 1948, and the University of Wisconsin, from which he retired in 1973. From 1945 to 1946, he was chief of the Punched Card Section of the Computing Branch of the Ballistics Research Laboratory of the Aberdeen Proving Grounds. From 1965 until his retirement he held a joint appointment at the University of Wisconsin and the Army Research Center in Madison (Schoenberg [2], back cover.)

In 1938, while proctoring a two-hour mechanics examination at Colby College (Schoenberg [3]), I.J. Schoenberg, starting out with the continuous mapping (5.1.2) from the Cantor set Γ onto $\mathcal{Q}$, came up with a new method for extending the definition of that mapping into $\mathcal{I}$ (Schoenberg [1]). His interpolation made it obvious that the function thus obtained is continuous on the entire interval $\mathcal{I}$ and a laborious proof such as the one in Theorem 5.4.1 can be avoided.

Schoenberg defined the components φ_{sc}, ψ_{sc} of a map $f_{sc} : \mathcal{I} \to \mathcal{Q}$ in terms of the even, two-periodic generating function

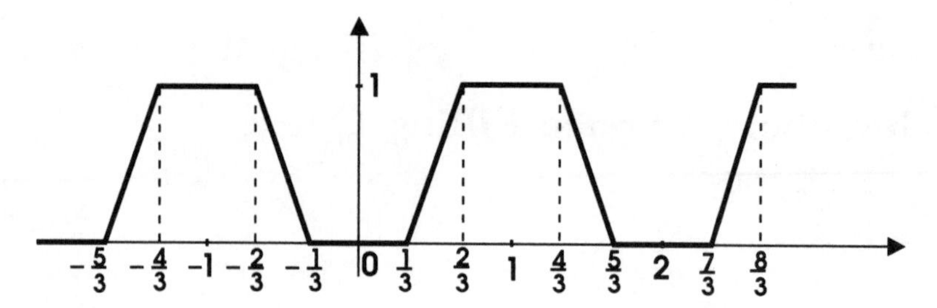

Fig. 7.1. Graph of the Generating Function p

$$p(t) = \left\{ \begin{array}{lll} 0 & \text{for} & 0 \le t \le \frac{1}{3} \\ 3t-1 & \text{for} & \frac{1}{3} \le t \le \frac{2}{3} \\ 1 & \text{for} & \frac{2}{3} \le t \le 1 \end{array} \right\}, \quad p(-t) = p(t), \quad p(t+2) = p(t) \quad (7.1.1)$$

(see also Fig. 7.1) as follows:

$$\varphi_{sc}(t) = \frac{1}{2}\sum_{k=0}^{\infty} p(3^{2k}t)/2^k, \psi_{sc}(t) = \frac{1}{2}\sum_{k=0}^{\infty} p(3^{2k+1}t)/2^k. \qquad (7.1.2)$$

The mapping that is defined in (7.1.2) is continuous. This may be seen as follows: The series in (7.1.2) are dominated by the geometric series $\sum_{n=0}^{\infty} \left(\frac{1}{2}\right)^n$ and are, therefore, uniformly convergent. The individual terms are continuous and, since a uniformly convergent series of continuous functions represents a continuous function (see Appendix A.2.4), φ_{sc}, ψ_{sc} are continuous. □

The mapping defined in (7.1.2) is surjective onto $\mathcal{Q}$. We will show that the restriction of f_{sc} to the Cantor set Γ is identical with the mapping in (5.1.2). Since the latter maps $\mathcal{I}$ onto $\mathcal{Q}$, surjectivity is assured.

Let $t \in \Gamma$:

$$t = 2b_1/3 + 2b_2/3^2 + 2b_3/3^3 + \cdots, b_j = 0 \text{ or } 1.$$

Then

$$\begin{aligned} 3^{2k}t &= 2b_1 3^{2k-1} + 2b_2 3^{2k-2} + \cdots + 2b_{2k} + 2b_{2k+1}/3 + 2b_{2k+2}/3^2 + \cdots \\ &= \text{even integer} + 2b_{2k+1}/3 + 2b_{2k+2}/3^2 + \cdots . \end{aligned}$$

From (7.1.1), and since

$$2b_{2k+2}3^2 + 2b_{2k+3}/3^3 + \cdots \le 1/3,$$

we have

$$p(3^{2k}t) = \begin{cases} 0 \text{ if } & b_{2k+1} = 0 \\ 1 \text{ if } & b_{2k+1} = 1 \end{cases}.$$

In other words, $p(3^{2k}t) = b_{2k+1}$. Hence,

$$\varphi_{sc}(t) = \frac{1}{2}\sum_{k=0}^{\infty} b_{2k+1}/2^k = 0_{\dot{2}}b_1b_3b_5\ldots .$$

In a similar manner, we obtain

$$\psi_{sc}(t) = \frac{1}{2}\sum_{k=0}^{\infty} b_{2k+2}/2^k = 0_{\dot{2}}b_2b_4b_6\ldots,$$

and we see that the restriction of f_{sc} to Γ is indeed identical with the mapping of (5.1.2). □

We summarize these results in

(7.1) Theorem. *The function* $f_{sc}: \mathcal{I} \to \mathcal{Q}$, *which is defined in (7.1.2), represents a space-filling curve. We call it the Schoenberg curve.*

7.2. The Nowhere Differentiability of the Schoenberg Curve

Since the restriction of f_{sc} to Γ is identical to the mapping (5.1.2), and since the mapping (5.1.2) is nowhere differentiable (see Theorem 5.1), it follows that the restriction of the Schoenberg curve to Γ is nowhere differentiable. The question now involves what happens on the rest of $\mathcal{I}$. We have seen in Section 5.4 that Lebesgue's space-filling curve is differentiable on Γ^c. It will develop that the Schoenberg curve is not. In fact, the Schoenberg curve is nowhere differentiable on $\mathcal{I}$. J.-B. Hiriart-Urruty states "*Le fait que* f_{sc} *ne soit nulle part differentiable est plus laborieux à prouver*" ("The fact that f_{sc} is nowhere differentiable is very laborious to prove"—Hiriart-Urruty [1], p. 230). The first proof did not appear until 1981. This proof, by J. Alsina, is indeed laborious (Alsina [1]). Schoenberg, a year later, published a proof (in Schoenberg [2]) that, for $0 < a < 1, b$ odd, and $ab > 1$, the function $\sum_{k=0}^{\infty} a^k E(b^k t)$, where E is the linear Euler spline that interpolates $\cos(\pi t)$ between consecutive integers, is nowhere differentiable. He does this by adapting Rudin's version of the proof of the nowhere differentiability of Weierstraß' first example of such a function (Rudin [1], pp. 125–127). He then deduces from this that $3\varphi_{sc}(t)+\psi_{sc}(t)$ is nowhere differentiable, whence the desired result follows. Again, the proof is laborious and lengthy. In 1986, we found an elementary proof, which we eventually published (Sagan [4]). This is the proof we will present here:

(7.2) Theorem. *The Schoenberg curve, as defined in (7.1.2), is nowhere differentiable.*

Proof. We have to distinguish three cases: $t = 0, t = 1$, and $t \in (0,1)$.

(i) $t = 0$. We choose $t = 1/9^n$. Since

$$p(9^k/9^n) = \begin{cases} 0 \text{ for } & k < n \\ 1 \text{ for } & k \geq n \end{cases},$$

we have

$$\varphi_{sc}(1/9^n) = \frac{1}{2}\sum_{k=n}^{\infty} 1/2^k = 1/2^n$$

and, hence,

$$[\varphi_{sc}(1/9) - \varphi_{sc}(0)]/(1/9^n) = (9/2)^n \longrightarrow \infty$$

as $n \longrightarrow \infty$, i.e., $\varphi'_{sc}(0)$ does not exist.

(ii) $t = 1$. Let $t_n = 1 - 1/9^n$ and proceed as before.

(iii) Let $t \in (0,1)$. Choose $k_n = [9^n t]$ and let $a_n = k_n/9^n$, $b_n = a_n + 1/9^n$. If n is sufficiently large, then $0 < a_n < t < b_n < 1$ and $\{a_n\} \longrightarrow t, \{b_n\} \longrightarrow t$ unless t has a finite eneadic representation in which case $a_n = t$ from a certain subscript n on. In the sequence of integers $\{k_n\}$, infinitely many are even or infinitely many are odd (or both). Let us assume that infinitely many are even, and let us denote this subsequence of even integers again by k_n, in order to avoid double subscripts. From (7.1.2),

$$\varphi_{sc}(a_n) = \frac{1}{2}\sum_{k=0}^{\infty} p(9^k k_n/9^n)/2^k, \varphi_{sc}(b_n) = \frac{1}{2}\sum_{k=0}^{\infty} p(9^k k_n/9^n + 9^k/9^n)/2^k.$$

Hence,

$$\varphi_{sc}(b_n) - \varphi_{sc}(a_n) = \frac{1}{2}\sum_{k=0}^{n-1}[p(9^k k_n/9^n + 9^k/9^n) - p(9^k k_n/9^n)]/2^k$$
$$+ \frac{1}{2}\sum_{k=n}^{\infty}[p(9^k k_n/9^n + 9^k/9^n) - p(9^k k_n/9^n)]/2^k \stackrel{\triangle}{=} S_1 + S_2. \qquad (7.2.1)$$

If $k < n$, then $9^k/9^n \leq 1/9$ and, in order to obtain a lower estimate for S_1, we assume the worst possible situation, where $9^k k_n/9^n + 9^k/9^n$ and $9^k k_n/9^n$ both lie in an interval where p descends with slope -3 (see Fig. 7.1). Then

$$p(9^k k_n/9^n + 9^k/9^n) - p(9^k k_n/9^n)] \geq -3(9^k/9^n).$$

Hence,

$$S_1 \geq -\frac{3}{9^n \cdot 2}\sum_{k=0}^{n-1}(9/2)^k = -\frac{3}{9^n \cdot 7}[(9/2)^n - 1]. \qquad (7.2.2)$$

If $k \geq n$, then $(9^k/9^n) \geq 1$, odd. Hence $(9^k/9^n)k_n$ is even and $9^k k_n/9^n + 9^k/9^n =$ even + odd = odd. Therefore,

$$S_2 = \frac{1}{2}\sum_{k=n}^{\infty}[p(\text{odd}) - p(\text{even})]/2^k = \frac{1}{2}\sum_{k=n}^{\infty}(1/2)^k = 1/2^n. \qquad (7.2.3)$$

From (7.2.1), (7.2.2), and (7.2.3) we obtain

$$[\varphi_{sc}(b_n) - \varphi_{sc}(a_n)]/(b_n - a_n) = 9^n(S_1 + S_2) \geq \frac{4}{7}\left(\frac{9}{2}\right)^n + \frac{3}{7} \longrightarrow \infty$$

as $n \to \infty$, i.e., $\varphi_{sc}'(t)$ does not exist (see Appendix A.2.2).

Suppose now that infinitely many elements in the sequence $\{k_n\}$ are odd. We reverse our strategy and seek an upper bound for the difference quotient. By similar reasoning,

$$[\varphi_{sc}(b_n) - \varphi_{sc}(a_n)]/(b_n - a_n) \leq -\frac{4}{7}\left(\frac{9}{2}\right)^n - \frac{3}{7} \longrightarrow -\infty \qquad (7.2.4)$$

as $n \longrightarrow \infty$.

Since $\psi_{sc}(t) = \varphi_{sc}(3t)$ by (7.1.2), ψ_{sc} is also nowhere differentiable. □

Corollary to Theorem 7.2. *The restriction of f_{sc} to Γ is nowhere differentiable and the restriction to Γ^c is also nowhere differentiable.*

Proof. We have seen in Section 7.1 that the restriction of f_{sc} to the Cantor set Γ is identical with the mapping (5.1.2), which, by Theorem 5.1, is nowhere differentiable. All that is left to prove is that the restriction of f_{sc} to Γ^c is nowhere differentiable. Since Γ^c, being a union of open sets, is open, and since φ_{sc} is nowhere differentiable in $\mathcal{I}$, there is, for every $t \in \Gamma^c$, a sequence $\{t_n\} \to t$ with $t_n \in \Gamma^c$ such that $\lim_{n\to\infty}[\varphi_{sc}(t) - \varphi_{sc}(t_n)]/(t - t_n)$ does not exist. The same argument applies to ψ_{sc}. □

(This corollary constitutes a complete answer to a question Schoenberg raised (Schoenberg [2], p. 146, remark 1).

7.3. Approximating Polygons

Consistent with our previous usage, and with the notion we introduced in Sagan [2], we call the polygonal line that joins the nodal points $f_{sc}(m/3^n), m = 0, 1, 2, \ldots, 3^n$, *the nth approximating polygon to the Schoenberg curve.* It is possible to find the exact coordinates of the nodal points by finitely many operations. This may be seen as follows. Let

$$t_{m,n} = m/3^n, n = 1, 2, 3, \ldots; m = 0, 1, 2, \ldots, 3^n.$$

From (7.1.1),

$$p(3^j t_{m,n}) = p(m) = m - 2[m/2] \text{ for all } j \geq n.$$

Hence,

$$\begin{aligned}
\varphi_{sc}(t_{m,n}) &= \frac{1}{2}\sum_{k=0}^{[n/2]} p(3^{2k}t_{m,n})/2^k + \frac{1}{2}\sum_{k=[n/2]+1}^{\infty} p(m)/2^k \\
&= \frac{1}{2}\sum_{k=0}^{[n/2]} p(3^{2k}t_{m,n})/2^k + (m-2[m/2])/2^{[n/2]+1} \\
\psi_{sc}(t_{m,n}) &= \frac{1}{2}\sum_{k=0}^{[n/2]} p(3^{2k+1}t_{m,n})/2^k + \frac{1}{2}\sum_{k=[n/2]+1}^{\infty} p(m)/2^k \\
&= \frac{1}{2}\sum_{k=0}^{[n/2]} p(3^{2k+1}t_{m,n})/2^k + (m-2[m/2])/2^{[n/2]+1}.
\end{aligned} \tag{7.3.1}$$

Now that we know $f_{sc}(t_{m,n})$, we define the nth approximating polygon to the Schoenberg curve as follows:

$$s_n(t) = 3^n(f_{sc}(t_{n,m+1}) - f_{sc}(t_{n,m}))(t - t_{n,m}) + f_{sc}(t_{n,m}) \tag{7.3.2}$$

for all $t \in [t_{m,n}, t_{m+1,n}], n = 1,2,3,\ldots; m = 0,1,2,\ldots,3^n - 1$. We obtain from (7.3.1) and (7.3.2) for the first component $\sigma_n(t)$ of $s_n(t)$:

$$\sigma_n(t) = \frac{1}{2}\sum_{k=0}^{\infty}[3^n(p(3^{2k}t_{m+1,n}) - p(3^{2k}t_{m,n}))(t - t_{m,n}) + p(3^{2k}t_{m,n})]/2^k.$$

We want to show that $\{\sigma_n(t)\}$ converges uniformly to $\varphi_{sc}(t)$ on $\mathcal{I}$. We note that $3^{2k}/3^n \leq 1/3$ whenever $2k < n$. Hence, $p(3^{2k}t)$ is represented on the interval $[t_{m,n}, t_{m+1,n}]$ by a straight line from its beginning point $(t_{m,n}, p(3^{2k}t_{m,n}))$ to its endpoint $(t_{m+1,n}, p(3^{2k}t_{m+1,n}))$,

$$p(3^{2k}t) = 3^n(p(3^{2k}t_{m+1,n}) - p(3^{2k}t_{m,n}))(t - t_{m,n}) + p(3^{2k}t_{m,n})$$

and, hence,

$$\begin{aligned}
|\sigma_n(t) - \varphi_{sc}(t)| &= \Bigg|\frac{1}{2}\sum_{k=[n/2]}^{\infty} 3^n(p(3^{2k}t_{m+1,n}) - p(3^{2k}t_{m,n}))(t - t_{m,n}) \\
&\quad + p(3^{2k}t_{m,n}) - p(3^{2k}t)]/2^k\Bigg| \\
&\leq \frac{1}{2}\sum_{k=[n/2]}^{\infty} [3^n|p(3^{2k}t_{m+1,n}) - p(3^{2k}t_{m,n})||t - t_{m,n}| \\
&\quad + |p(3^{2k}t_{m,n})| + |p(3^{2k}t)|]/2^k.
\end{aligned}$$

Since $0 \leq p(t) \leq 1$, we have $|p(3^{2k}t_{m+1,n}) - p(3^{2k}t_{m,n})| \leq 1$ and since $t \in [t_{m,n}, t_{m+1,n}]$, we have $|t - t_{m,n}| \leq 1/3^n$. Hence,

$$|\sigma_n(t) - \varphi_{sc}(t)| \leq \frac{1}{2} \sum_{k=[n/2]}^{\infty} 3/2^k = 3/2^{[n/2]} \tag{7.3.3}$$

and we see that $\{\sigma_n(t)\} \to \varphi_{sc}(t)$ uniformly on $\mathcal{I}$. By the same reasoning one obtains a similar result for the second component of $s_n(t)$, and we have indeed $\{s_n(t)\} \to f_{sc}(t)$ uniformly on $\mathcal{I}$. The first, second, third, and fourth approximating polygons are depicted in Fig. 7.3.1, which was generated by Computer Program 4 in Appendix A.1.4.

By contrast, Alsina [1] considers the following sequence a_n of polygons with components α_n, β_n:

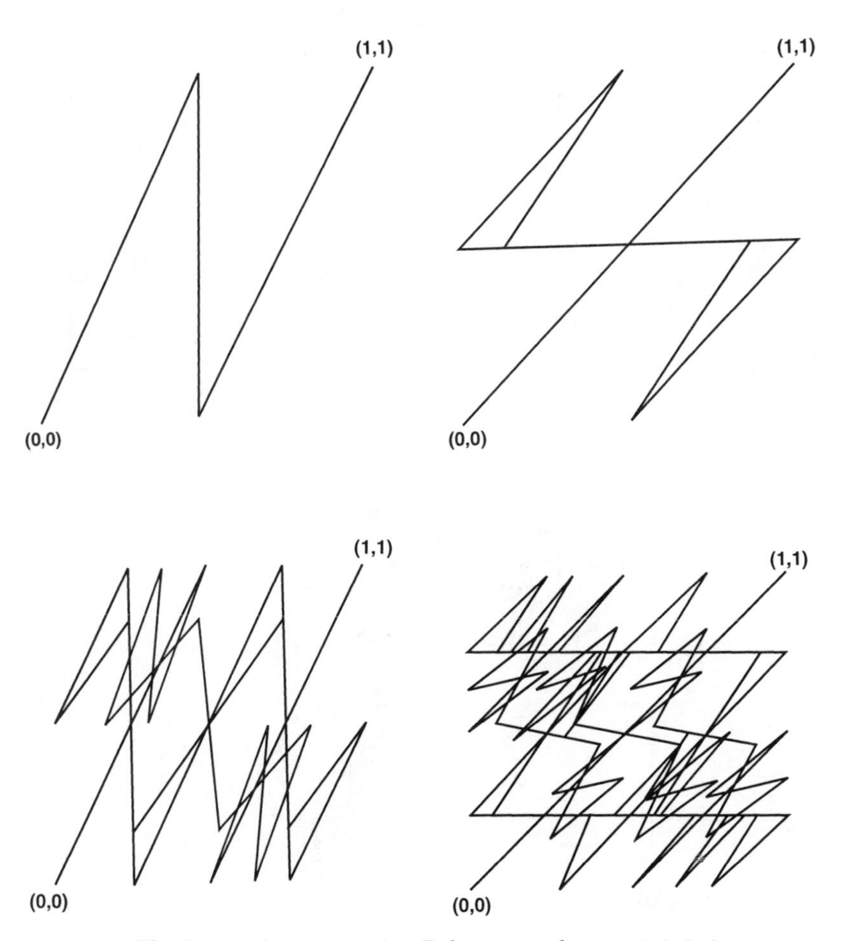

Fig. 7.3.1. Approximating Polygons s_n for $n = 1, 2, 3, 4$

$$\alpha_n(t) = \frac{1}{2}\sum_{k=0}^{n-1} p(3^{2k}t)/2^k, \beta_n(t) = \frac{1}{2}\sum_{k=0}^{n-1} p(3^{2k+1}t)/2^k. \qquad (7.3.4)$$

From (7.1.2),

$$|\varphi_{sc}(t) - \alpha_n(t)| \leq 1/2^n, |\psi_{sc}(t) - \beta_n(t)| \leq 1/2^n, \qquad (7.3.5)$$

and we see that the polygons in (7.3.4) also approximate the Schoenberg curve uniformly. It is therefore justified to also call them *approximating polygons*. The approximating polygons a_n are depicted in Fig. 7.3.2 for $n = 1, 2, 3$. These polygons were generated by Computer Program 5 in Appendix A.1.4.

Lebesgue's space-filling curve was obtained (in Chapter 5) from the mapping (5.1.2) by linear interpolation, meaning that the joins on

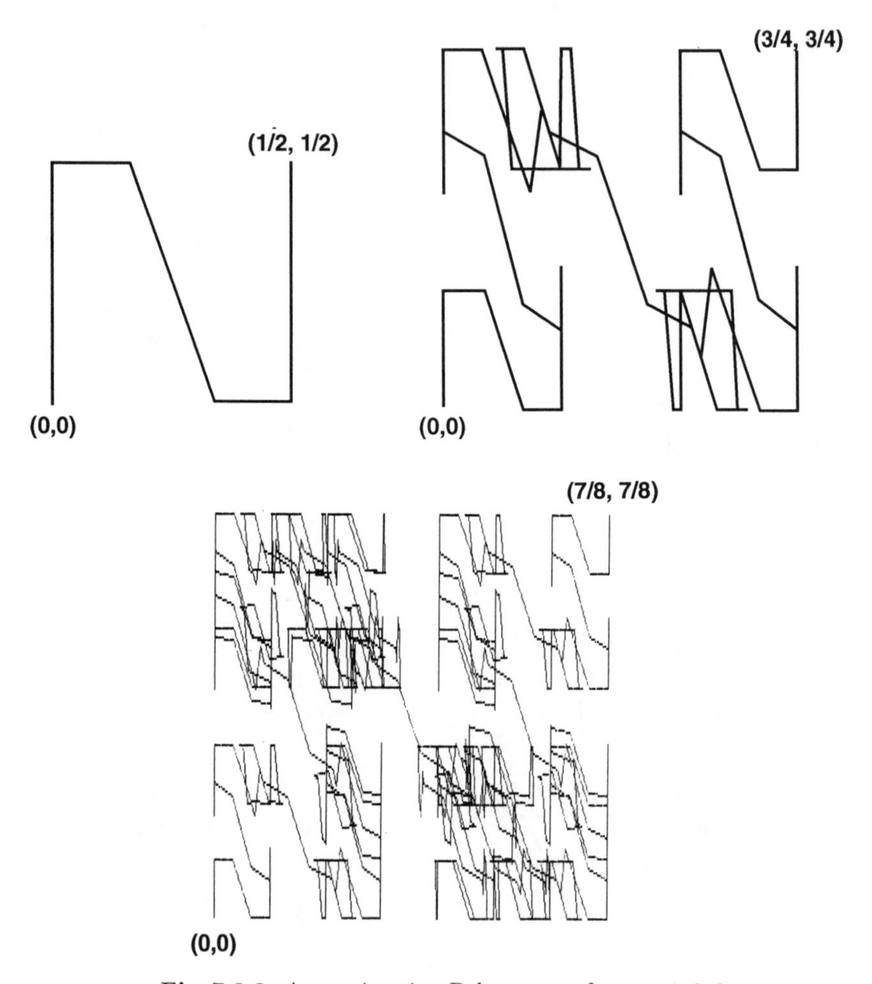

Fig. 7.3.2. Approximating Polygons a_n for $n = 1, 2, 3$

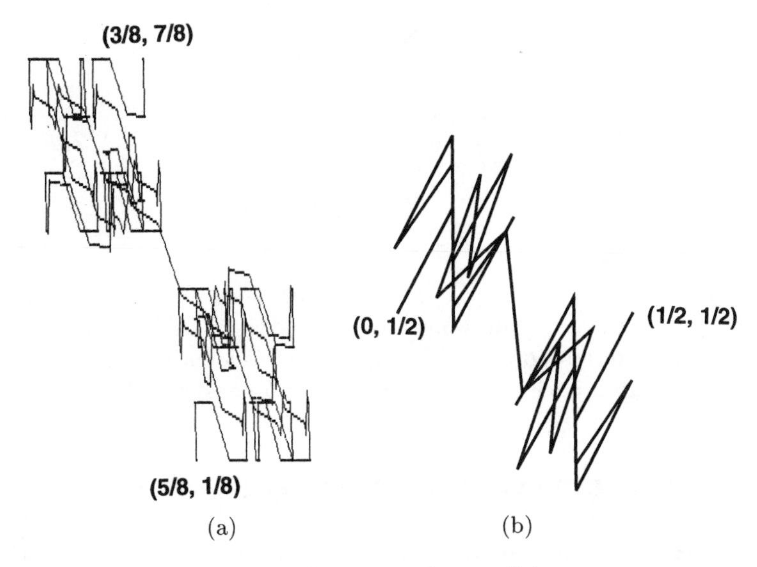

Fig. 7.3.3. Approximations to Joins

the intervals that make up Γ^c are straight lines (see also Fig. 5.5.3). Since Schoenberg's curve is nowhere differentiable, one can expect very complicated joins and this is exactly what happens. In Fig. 7.3.3(a) we exhibit an a_3-approximation to the join in the interval $[1/3, 2/3]$ and in Fig. 7.3.3(b) an s_5-approximation to the join on the interval $[1/9, 2/9]$. Note that the s_5-approximation to the join on $[1/9, 2/9]$ joins the nodal point $(1/2, 1/2)$ to the nodal point $(0, 1/2)$, the same as the join. However, the a_3-approximation to the join on $[1/3, 2/3]$ does not join the nodal point $(1/2, 1)$ to the nodal point $(1/2, 0)$ as the join does, but rather an approximation $(3/8, 7/8)$ of the first nodal point to an approximation $(5/8, 1/8)$ of the second nodal point.

Before closing this section, let us point to a recent paper by E. Hlawka on a class of approximating polygons to the (Schoenberg) Peano curve (Hlawka [1]), where he uses the theory of uniform distribution to construct as many points of the (two-dimensional) Schoenberg curve as needed, shows how this may be used to find approximate solutions to equations, and develops a new theory of uniform distribution, using a singular measure, which may be applied to Brownian motion.

7.4. A Three-Dimensional Schoenberg Curve

Schoenberg defines a three-dimensional curve in terms of the generating function p (see Fig. 7.1.1) as follows:

$$\xi = \frac{1}{2}\sum_{k=0}^{\infty} p(3^{3k}t)/2^k$$
$$\eta = \frac{1}{2}\sum_{k=0}^{\infty} p(3^{3k+1}t)/2^k \tag{7.4.1}$$
$$\zeta = \frac{1}{2}\sum_{k=0}^{\infty} p(3^{3k+2}t)/2^k$$

for $0 \le t \le 1$ (Schoenberg [2]). The mapping is continuous since the series converge uniformly and all terms are continuous. As in Section 7.1, one can show that the restriction of this mapping to Γ is identical with (5.1.3) and, hence, the map is surjective onto $\mathcal{W}$. It is also nowhere differentiable.

7.5. An $\aleph_0$-Dimensional Schoenberg Curve

Sierpiński credits the Japanese mathematician Kiyoshi Iséki with the following construction of an $\aleph_0$-dimensional Schoenberg curve (Sierpiński [10], p. 78):

Let p be defined as in (7.1.1) and

$$\varphi_n(t) = \sum_{k=1}^{\infty} p(3^{2^{n-1}(2k-1)-1}t)/2^k \tag{7.5.1}$$

for $n = 1, 2, 3, \ldots$. These functions, being represented by uniformly convergent series of continuous functions, are continuous. Since $0 \le p(t) \le 1$, we have $0 \le \varphi_n(t) \le 1$.

In order to show that

$$\xi_n = \varphi_n(t), t \in \mathcal{I}, n = 1, 2, 3, \ldots$$

represents an $\aleph_0$-dimensional space-filling curve, we only need to demonstrate that for any point $(\alpha_1, \alpha_2, \alpha_3, \ldots) \in \mathcal{I} \times \mathcal{I} \times \mathcal{I} \times \ldots$ there is a $t \in \mathcal{I}$ such that

$$\varphi_n(t) = \alpha_n \quad \text{for } n = 1, 2, 3, \ldots .$$

Let $\alpha_n = 0_{\dot{2}}\alpha_n^{(1)}\alpha_n^{(2)}\alpha_n^{(3)}\ldots, \alpha_n^{(k)} = 0$ or 1, and let $t_{2^{n-1}(2k-1)} = \alpha_n^{(k)}$.

We will demonstrate that

$$t = 2\sum_{m=1}^{\infty} t_m/3^m$$

is the value of t we are looking for. We have for $q = 0, 1, 2, 3, \ldots$

$$3^q t = 2\sum_{m=1}^{\infty} 3^q t_m/3^m = 2N(q) + (2/3)t_{q+1} + (2/3^2)t_{q+2} + \cdots, \tag{7.5.2}$$

where $N(q)$ is some integer, depending on q.

If $t_{q+1} = 0$, we have from (7.5.2) that

$$0 \le 3^q t - 2N(q) \le (2/3^2) + (2/3^3) + \cdots = 1/3$$

and, hence, by (7.1.1),

$$p(3^q t) = p(3^q t - 2N(q)) = 0 = t_{q+1}.$$

If $t_{q+1} = 1$, then, again from (7.5.2),

$$3^q t - 2N(q) = (2/3) + (2/3^2)t_{q+2} + \cdots$$

and, hence,

$$\begin{aligned} 2/3 = 3^q t - 2N(q) - (2/3^2)t_{p+2} + \cdots &\le 3^q t - 2N(q) \\ \le 2/3 + (2/3^2) + \cdots = 1. \end{aligned}$$

From (7.1.1),

$$p(3^q t) = 1 = t_{q+1}.$$

So, we have

$$p(3^q t) = t_{q+1} \quad \text{for } q = 0, 1, 2, 3, \ldots .$$

Therefore

$$\varphi_n(t) = \sum_{k=1}^{\infty} t_{2^{n-1}(2k-1)}/2^k = \alpha_n, \quad \text{for } n = 1, 2, 3, \ldots . \square$$

The above provides a constructive proof that the $\aleph_0$-dimensional unit cube has cardinality $\mathfrak{c}$. (See also Section 6.9.)

7.6. Problems

1. Use the defining equations (7.1.2) of Schoenberg's space-filling curve to find the values of f_{sc} at $t = 1/9, 2/9, 1/3, 2/3, 5/9$, and check your result against (5.1.2).
2. Show: if $3\varphi_{sc} + \psi_{sc}$ is nowhere differentiable on $\mathcal{I}$, then φ_{sc} and ψ_{sc} are nowhere differentiable on $\mathcal{I}$.
3. Prove that the Schoenberg curve is not differentiable at $t = 1$.
4. With $k_n = [9^n t]$, $t \in (0, 1)$, and $a_n = k_n/9^n$, $b_n = a_n + 1/9^n$, show that $0 < a_n < t < b_n < 1$ for sufficiently large n and that $\{a_n\} \to t$ and $\{b_n\} \to t$ as $n \to \infty$, unless t has a finite eneadic representation, in which case $a_n = t$ from a certain subscript n on.
5. Carry out the details in establishing the validity of (7.2.4).
6. Show that $p(m) = m - 2[m/2]$ for all integers m, where p is defined in (7.1.1).

7. It would appear from the estimates in (7.3.3) and (7.3.5) that the a_n-polygons approach the Schoenberg curve faster than the s_n-polygons. Is this true? Justify your answer.
8. Show that
$$\left.\begin{aligned} x &= (t+1-\psi_{sc}(t))/2 \\ y &= 1-\varphi_{sc}(t) \end{aligned}\right\}, \quad t \in [0,1]$$
represents the Schoenberg curve on the interval $[1/3, 2/3]$.
9. Show that
$$\left.\begin{aligned} x &= 1-\varphi_{sc}(t)/2 \\ y &= (t+1-\psi_{sc}(t))/2 \end{aligned}\right\}, t \in [0,1]$$
represents the Schoenberg curve on the interval $[7/9, 8/9]$.
10. Find representations similar to the ones in Problems 8 and 9 on the intervals $[1/9, 2/9]$, $[1/27, 2/27]$.
11. Is the Schoenberg curve space-filling on the intervals $[1/3, 2/3]$, $[1/9, 2/9]$, $[7/9, 8/9]$, etc.?
12. Show that the restriction of the three-dimensional Schoenberg curve (7.4.1) to Γ is identical with the mapping (5.1.3).
13. Prove that the three-dimensional Schoenberg curve in (7.4.1) is nowhere differentiable.
14. With $\varphi_n(t)$ as defined in (7.5.1), find values of t for which (a) $\varphi_n(t) = 1$ for all $n = 1, 2, 3, \ldots$, (b) $\varphi_n(t) = 0$ for all $n = 1, 2, 3, \ldots$, (c) $\varphi_j(t) = 1, \varphi_n(t) = 0$ for all $n \neq j$, (d) $\varphi_j(t) = 0$, $\varphi_n(t) = 1$ for all $n \neq j$.
15. Show that $2^{n-1}(2k-1) = m$ has integer solutions n, k for each $m = 1, 2, 3, \ldots$.

Chapter 8

Jordan Curves of Positive Lebesgue Measure

8.1. Jordan Curves

We know from Netto's theorem (Theorem 6.4) that plane curves that do not cross (or touch) themselves cannot be space-filling. We will show in this chapter that it is still possible for such curves to have a positive two-dimensional Lebesgue measure.

Many authors refer to curves without multiple points as *Jordan Curves*, although Camille Jordan himself defined a curve in his *Course d'Analyse* (Jordan [1], p. 90) as we did in Definition 1.3.2 and held to that definition through three editions. (A reprinting of the third edition was reissued as late as 1959.) He proceeded to prove that a closed curve without multiple points partitions the plane into an inside and an outside. This is now known as the Jordan Curve theorem and it seems that this is the reason that the term *Jordan curves* became attached to continuous curves without multiple points. Here is what O. Veblen had to say about Jordan's proof: "It assumes the theorem without proof in the important special case of a simple polygon and of the argument from that point on, one must admit at least that all details are not given" (Veblen [2], p. 83). As W.F. Osgood observed, for the polygons alone, the theorem is not evident, because one can only have a clear conception of a very restricted class of polygons ("*Allein für die Polygone ist der Satz schon nicht mehr anschaulich, denn man kann sich ja nur von einer äußerst beschränkten Klasse von Polygonen eine deutliche Vorstellung machen*"—Osgood [3], p. 171). In 1904, Veblen published a paper on a system of axioms for geometry (Veblen [1]), which, incidentally, is equivalent to Hilbert's axioms of order and connection (Hilbert [2]), and proved a year later (Veblen [2]), on the basis of these axioms, that Jordan's theorem is true for closed polygons without multiple points. Hahn, in turn, found Veblen's proof to be "not binding" and proceeded to give his own proof (Hahn [8]). In any event, the theorem is now considered to be proved.

On p. 102 of his *Cours d'Analyse*, Jordan proves that a rectifiable curve has two-dimensional content zero. His argument was simple and ingenious. Suppose the curve C has length L. Partition the plane with a square grid of sidelength L/n and the curve into n parts of equal length

L/n. Each part passes through, at most, four squares of the grid and, hence, $J_2(\mathcal{C}) < 4L^2/n \longrightarrow 0$ as $n \longrightarrow \infty$. Then, something strange happened: One year after the publication of the first edition of the *Cours d'Analyse* in 1893, he asked the mathematicians of the (civilized) world in the *l'Intermediaire des Mathematiciens* (loosely translated: The mathematicians' bulletin board), Vol. 1 (1894), p. 23 (Jordan [2]), if the area of a continuous curve was not indeterminate (*"Pourrait-on signaler une courbe $x = f(t)$, $y = \varphi(t)$ (f et φ étant des fonctions continues) dont l'aire fût indéterminée ?")*. Since he used the same term "indeterminée" as in conjunction with forms 0/0, it is not entirely clear what he meant by this. Anyway, it is astounding that Jordan was not yet aware of the existence of space-filling curves especially since Peano's and Hilbert's papers on the subject had already been published in 1890 and 1891. In 1896, Peano answered, in Vol. 3 of the *l'Intermediaire des Mathematiciens*, with a reference to his own space-filling curve (Peano [2], p. 39).

It appears that it was A. Hurwitz who, for the first time, injected injectivity into the definition of curves. In his programmatic keynote address to the First International Congress of Mathematicians (Hurwitz [1]), which dealt among others with contours admissible for Cauchy's integral theorem, he introduced on p. 102 the concept of a simple closed continuous curve as the image of a continuous 1-1 map from the border of a square onto the curve, pointing out that the concept of a continuous curve does not correspond to one's perception of a curve in view of the examples by Peano and Hilbert. (It is noteworthy that, at that time, Hurwitz still believed in the validity of Jordan's proof of Jordan's curve theorem. Incidentally, Jordan was not present at that Congress but Peano was.) We will adapt Hurwitz's notion to the case of simple curves that are not closed and give the following:

(8.1) Definition. *$\mathcal{C}$ is called a Jordan curve (or simple continuous curve) if it is the image of a continuous injective map $f : \mathcal{I} \longrightarrow \mathbb{E}^n$.*

We will present in this chapter some examples of Jordan curves, represented by $f : \mathcal{I} \rightarrow \mathcal{Q}$ (or $\mathcal{T}$), with f continuous and injective, for which $\Lambda_2(f_*(\mathcal{I})) > 0$. In fact, we will produce, for any given number between 0 and 1, *a* Jordan curve, the two-dimensional Lebesgue measure of which is equal to that number.

8.2. Osgood's Jordan Curves of Positive Measure

While the problem of finding Jordan curves of positive two-dimensional Lebesgue measure generated considerable activity and produced ingenious solutions early this century, it appears that it was all but forgotten as the century wore on because in 1982 there appeared an inquiry by F. Burton

Jones in the *American Mathematical Monthly* (Vol. 89, No. 10, Dec. 1982, p. 756, E2975) regarding the existence of Jordan curves of positive two-dimensional Lebesgue measure. Four years later, a response came forth (*American Mathematical Monthly*, Vol. 93, No. 7, July-August 1986, p. 569) referring the inquisitor to Gelbaum and Olmsted [1]. There, Gelbaum and Olmsted make reference to W.F. Osgood as the discoverer of the first such curve. Osgood's curves will be the subject of this section.

William Fogg Osgood (1864–1943) was born in Boston and died in Belmont, Massachusetts. He studied at Harvard under B.O. Peirce and F.N. Cole, and graduated second in his class in 1896. After an additional year of graduate studies, he received a Master's degree in 1887. The same year he went to the University of Göttingen, where he studied under Felix Klein, and from there to the University of Erlangen, where he received his doctor's degree in 1889. He returned to Harvard in 1890, and remained on the Harvard faculty until 1933. Until 1891, when Maxime Bôcher joined the department, Osgood was the only research-oriented Mathematician on the Harvard faculty. After his retirement in 1933, he taught for two years at the National University of Peiping. While at Peiping, he wrote to the Harvard College secretary of the class of 1886, in 1934 that "the best students here rank as high and are as promising as the best students at Harvard." Osgood is known for his contributions to the theory of functions of a complex variable, differential equations, and the calculus of variations. (For more details, see Walsh [1].)

In his pioneering paper, Osgood states that he was "led to the construction of the curve" (Jordan curve of positive two-dimensional Lebesgue measure) "through the consideration of some of the possibilities for the boundary of a simply connected region whose boundary points can all be approached along curves lying wholly in the region." (Osgood [2]. For simply connected regions for which this is not true, see Osgood [1].)

As the reader will see, the construction was, without a doubt, inspired by the geometric generation of the Peano curve, which appeared in print three years earlier (Moore [1]).

Osgood's construction of a family of Jordan curves with positive two-dimensional Lebesgue measure consists in the successive removal of open, grate-shaped regions (that are made up of horizontal and vertical bars) from squares, starting out with the unit square, and proceeding as indicated in Fig. 8.2.1 for the first two steps. The shaded closed squares are what is left after each step. The squares are then connected by "joins" as indicated

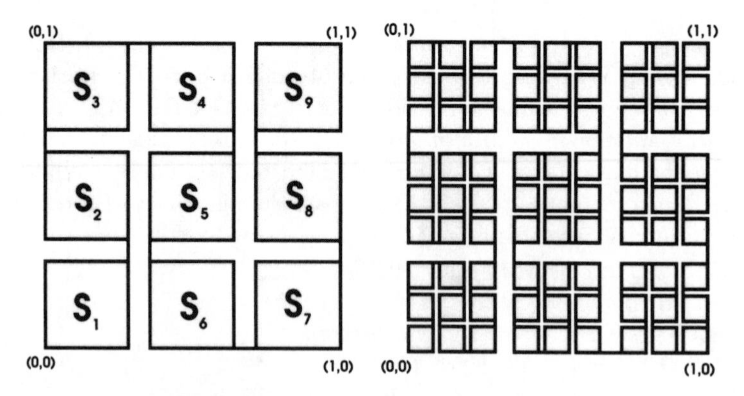

Fig. 8.2.1. Osgood's construction

by the bold line segments in Fig. 8.2.1. In order to construct a curve of two-dimensional Lebesgue measure λ for any $\lambda \in (0,1)$, we choose the grate-shaped regions so that the sum of their areas is $1-\lambda$. This may be accomplished as follows: At the first step, we choose the width of the bars to be $(1-\sqrt{\lambda})/4$. At the next step, $(1-\sqrt{\lambda})/24$, and at the next step, $(1-\sqrt{\lambda})/144$, etc....

Let $\mathcal{A}_n(\lambda)$ denote the region that consists of the 9^n closed squares that are left over after the nth step and the 9^{n-1} joins. Then, $J_2(\mathcal{A}_n) = [1-(1-\sqrt{\lambda})(1-1/2^n)]^2 = [\sqrt{\lambda}+(1-\sqrt{\lambda})/2^n]^2$ and, hence, the set $\mathcal{C}(\lambda) = \bigcap_{n=1}^{\infty} \mathcal{A}_n(\lambda)$ that is obtained after infinitely many steps has two-dimensional Lebesgue measure $\Lambda_2(\mathcal{C}(\lambda)) = \lim_{n\to\infty} J_2(\mathcal{A}_n(\lambda)) = \lambda > 0$ (see Appendix A.2.5).

In order to show that $\mathcal{C}(\lambda)$ is a Jordan curve, we will first parametrize it. To do so, we construct a Cantor-type set Γ_8 as follows: We remove from the interval $\mathcal{I}$ the eight open intervals $((2k-1)/17, 2k/17), k = 1,2,3,\ldots,8$, to be left with the nine closed intervals $[2k/17,(2k+1)/17], k = 0,1,2,\ldots,8$, of length $1/17$ each. We call the union of these nine closed intervals Γ_8^1. Next, we remove from each of the remaining closed intervals eight open intervals of length $1/17^2$ each, to be left with a total of 81 closed intervals of length $1/17^2$ each, the union of which we call Γ_8^2, etc.... $\Gamma_8 = \bigcap_{j=1}^{\infty} \Gamma_8^j(\lambda)$ is a Cantor-type set, each point of which is an accumulation point: a right accumulation point if it is a left endpoint of one of the removed intervals, a left accumulation point if it is the right endpoint of a removed interval, or a two-sided accumulation point. We now parametrize $\mathcal{C}(\lambda)$ as follows: We map $((2k-1)/17, 2k/17)$ linearly onto the join (without endpoints) from S_k to S_{k+1}, $k = 1,2,3,\ldots,8$ (see Fig. 8.2.1). The remaining closed intervals $[2k/17,(2k+1)/17]$ are mapped into the squares S_{k+1}, $k = 0,1,2,\ldots,8$. We repeat the process within each of these squares, and then within each of the squares within the squares, and continue the process ad infinitum. That this mapping is continuous can be shown by adapting the proof of

Theorem 5.4.1 (which deals with the continuity of Lebesgue's space-filling curve) to the new circumstances and is left to the reader. A simpler method for demonstrating continuity is to consider approximating polygons that are pieced together from the joins and the diagonals of the squares that connect the entry points into the squares to the exit points, as we have indicated in Fig. 8.2.2. The sequence of these approximating polygons converges uniformly to Osgood's curve, which is, therefore, continuous.

That the mapping is also injective may be seen as follows:

Any point on $\mathcal{C}(\lambda)$ either lies on a join or does not lie on a join. If it does, then, by construction, there is a unique preimage in one of the open intervals that have been removed in constructing Γ_8. If it does not, then it lies in one of the squares S_j, and, in turn, one of the subsquares of S_j, and, in turn, in one of the subsquares of the subsquare, etc.... The point is that at no time can it lie in two squares of the same partition. Therefore, the point is defined by a unique sequence of nested closed squares that shrink into a point which, in turn, corresponds to a unique sequence of nested closed intervals (the first of which belongs to Γ_8^1, the second of which belongs to Γ_8^2, etc. ...) that shrink to a point and defines a unique element in Γ_8.

We summarize our result:

(8.2) Theorem. *For any given* $\lambda \in (0,1), \mathcal{C}(\lambda) = \bigcap_{j=1}^{\infty} \mathcal{A}_j(\lambda)$ *is the trace of a Jordan curve and* $\Lambda_2(\mathcal{C}(\lambda)) = \lambda$*, whereby* $\mathcal{A}_n(\lambda)$ *denotes the union of squares (with total area* $J_2(\mathcal{A}_n(\lambda)) = (\sqrt{\lambda} + (1-\sqrt{\lambda})/2^n)^2)$ *that are left after* n *consecutive removals of grate-shaped regions (the bars of which are* $(1-\sqrt{\lambda})/4, (1-\sqrt{\lambda})/24, (1-\sqrt{\lambda})/144, \ldots$ *wide) and joins, as indicated in Fig. 8.2.1 for the first two steps.*

Note that this process yields a Jordan curve with a two-dimensional Lebesgue measure that is as close to 1 (the area of the containing square)

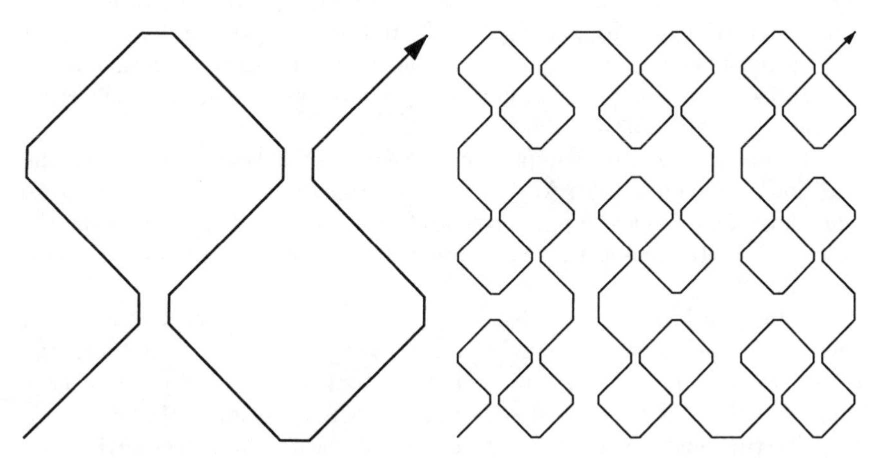

Fig. 8.2.2. Approximating Polygons for Osgood's Curve

as one pleases and, on the other hand, one of a two-dimensional Lebesgue measure as close to zero as one pleases. If one chooses $\lambda = 0$, one obtains a Jordan curve of two-dimensional Lebesgue measure zero. If one lets $\lambda \longrightarrow 1$, then the corresponding Jordan curves will tend towards the Peano curve (compare Fig. 8.2.1 with Fig. 3.3.2). The injectivity will be lost because, in the limit, the squares will touch and certain points on the curve will belong to more than one square and, hence, lead to more than one preimage. (For an elaborate description of Osgood's curve, see Beck, Bleicher, and Crowe [1], pp. 201–209.)

Jordan curves in $\mathbb{E}^n$ with positive n-dimensional Lebesgue measure are often referred to as *Osgood curves*.

8.3. The Osgood Curves of Sierpiński and Knopp

Knopp leveled some justifiable criticism at Osgood's construction (Knopp [1], p. 109, footnote 2). Because of the presence of the "joins", not every part of Osgood's curve is, by itself, an Osgood curve, and this is what Knopp criticized. He then proceeded to construct a curve without this shortcoming. But, prior to this, Sierpiński had already obtained an Osgood curve with the property that each part of it is, again, an Osgood curve. Although Knopp dismisses (in the same footnote) Sierpiński's construction as too complicated, we will give a brief outline because it was Sierpiński who "broke the ice" by taking the essential step that eliminated the need for joins.

Sierpiński started out with a right isosceles triangle $\mathcal{T}$, the hypotenuse of which has a length ≤ 1. He then removed an open rectangle of area $[J_2(\mathcal{T})]^2$ as indicated in Fig. 8.3.1(a), to be left with the three shaded closed triangles $\mathcal{T}_0, \mathcal{T}_1, \mathcal{T}_2$. Next, he removed from each of the shaded triangles open rectangles of areas $[J_2(\mathcal{T}_0)]^2, [J_2(\mathcal{T}_1)]^2, [J_2(\mathcal{T}_2)]^2$ to be left with the nine closed triangles $\mathcal{T}_{00}, \mathcal{T}_{01}, \mathcal{T}_{02}, \ldots, \mathcal{T}_{22}$ in Fig. 8.3.1(b), etc.... (The next step is illustrated in Fig. 8.3.1(c).) The intersection of all these shaded sets turns out to be a Jordan curve of positive two-dimensional Lebesgue measure (Sierpiński [3]).

Besides being "too complicated", Sierpiński's Osgood curve has the additional drawback that, for fixed initial triangle, the Lebesgue measure cannot be regulated as in the case of Osgood's example. Note, however, how the use of triangles that hang together at vertices eliminates the need for "joins."

Knopp picked up that idea but, instead of eliminating rectangles at each step, he eliminated triangles and came up with the following simple construction that allowed him to obtain Jordan curves of positive two-dimensional Lebesgue measure λ for any $\lambda \in (0, 1)$ (Knopp [1]).

Starting out with a triangle $\mathcal{T}$ (it need not be a right isosceles triangle), he removed an open triangle of area $r_1 J_2(\mathcal{T})$ where $r_1 \in (0, 1)$, to be left

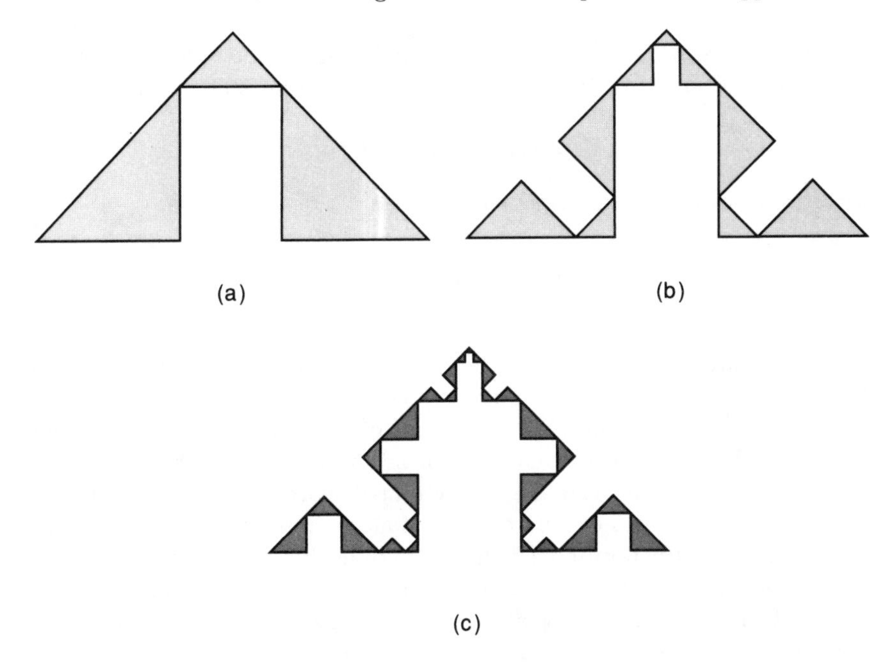

Fig. 8.3.1. Generation of Sierpiński's Osgood curve

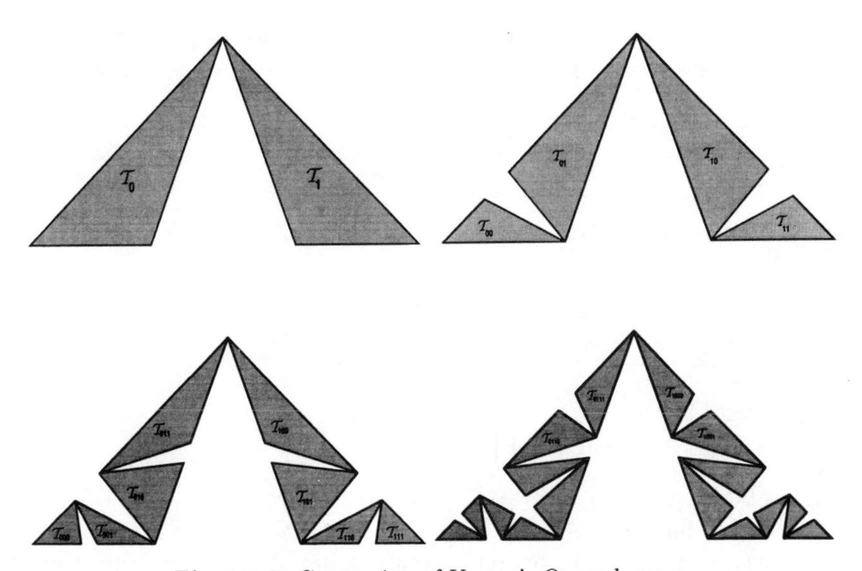

Fig. 8.3.2. Generation of Knopp's Osgood curve

with two closed triangles $\mathcal{T}_0, \mathcal{T}_1$ of a combined area $J_2(\mathcal{T})(1 - r_1)$. (See Fig. 8.3.2). From $\mathcal{T}_0$ he removed an open triangle of area $r_2 J_2(\mathcal{T}_0)$, and from $\mathcal{T}_1$ an open triangle of area $r_2 J_2(\mathcal{T}_1)$ for some $r_2 \in (0, 1)$ to be left with four closed triangles $\mathcal{T}_{00}, \mathcal{T}_{01}, \mathcal{T}_{10}, \mathcal{T}_{11}$ with a combined area of $J_2(\mathcal{T})(1-r_1)(1-r_2)$, and continued this process ad infinitum. In the limit, he obtained a point set

$$\mathcal{C} = (\mathcal{T}_0 \cup \mathcal{T}_1) \cap (\mathcal{T}_{00} \cup \mathcal{T}_{01} \cup \mathcal{T}_{10} \cup \mathcal{T}_{11}) \cap \ldots$$

with two-dimensional Lebesgue measure

$$\Lambda_2(\mathcal{C}) = J_2(\mathcal{T}) \prod_{j=1}^{\infty} (1 - r_j). \tag{8.3.1}$$

If the r_j are chosen so that $\sum_{j=1}^{\infty} r_j$ converges, then $\Lambda_2(\mathcal{C}) > 0$ (Appendix A.2.7). If, at each step, the triangles that are to be removed are placed judiciously, then all dimensions of the remaining triangles tend to zero, and the triangles themselves shrink into points. As Knopp noted, this is guaranteed, for example, if one starts out with an obtuse triangle and, at each step, cuts out triangles so that the remaining ones are again obtuse but with the obtuse angle not exceeding that of the initial triangle. If α is the smallest angle in the initial triangle, and c is the length of its base (side opposite the obtuse angle), then all angles of all triangles remain greater than or equal to α, and the bases of all triangles at the nth step of the construction cannot exceed $c \cos^n \alpha$, which tends to 0 as n tends to ∞. (Knopp [1]).

If we choose as initial triangle $\mathcal{T}$ a right isosceles triangle with a base of length 2 (and hence, $J_2(\mathcal{T}) = 1$), and $r_j = r^2/j^2$, for $r \in (0,1)$, we obtain

$$\Lambda_2(\mathcal{C}) = \prod_{j=1}^{\infty} (1 - r^2/j^2).$$

From Weierstraß' factorization theorem (Osgood [3], p. 555),

$$\sin \pi z = \pi z \prod_{j=1}^{\infty} (1 - z^2/j^2).$$

Hence,

$$\Lambda_2(\mathcal{C}) = \prod_{j=1}^{\infty} (1 - r^2/j^2) = \frac{\sin(\pi r)}{\pi r}. \tag{8.3.2}$$

Observe that for any $\lambda \in (0,1)$, $\frac{\sin(\pi r)}{\pi r} = \lambda$ has a solution $r \in (0,1)$. Hence, by judicious choice of r, one can obtain a set $\mathcal{C}(\lambda)$ of two-dimensional Lebesgue measure λ for any $\lambda \in (0,1)$.

We will now parametrize $\mathcal{C}$ and demonstrate that it is a Jordan curve: We construct a mapping $f : \mathcal{I} \longrightarrow \mathcal{C}$ by mapping the interval $[0, 1/2]$ into $\mathcal{T}_0$ and $[1/2, 1]$ into $\mathcal{T}_1$, with $t = 1/2$ going into the point that $\mathcal{T}_0, \mathcal{T}_1$ have in common, such that $[0, 1/4]$ goes into $\mathcal{T}_{00}$, $[1/4, 1/2]$ into $\mathcal{T}_{01}$, $[1/2, 3/4]$ into $\mathcal{T}_{10}$, and $[3/4, 1]$ into $\mathcal{T}_{11}$, with $1/4$ going into the point that is common to

$\mathcal{T}_{00}$ and $\mathcal{T}_{01}$, 1/2 into the point common to $\mathcal{T}_{01}$ and $\mathcal{T}_{10}$, and 3/4 into the point that is common to $\mathcal{T}_{10}$ and $\mathcal{T}_{11}$, such that... etc....

If we represent every $t \in (0,1]$ by an infinite binary:

$$t = 0_{\dot{2}} b_1 b_2 b_3 \ldots$$

(every finite binary $0_{\dot{2}} b_1 b_2 b_3 \ldots b_n 1$ being replaced by the infinite binary, $0_{\dot{2}} b_1 b_2 b_3 \ldots b_n 0\bar{1}$), then, by the above mapping, $f(t)$ lies in $\mathcal{T}_{b_1}$ and in $\mathcal{T}_{b_1 b_2}$ and in $\mathcal{T}_{b_1 b_2 b_3}$, etc.... Since $\mathcal{T}_{b_1} \supset \mathcal{T}_{b_1 b_2} \supset \mathcal{T}_{b_1 b_2 b_3} \supset \mathcal{T}_{b_1 b_2 b_3 b_4} \supset \ldots$ is a nested sequence of closed triangles that shrink into points,

$$\mathcal{T}_{b_1} \cap \mathcal{T}_{b_1 b_2} \cap \mathcal{T}_{b_1 b_2 b_3} \cap \mathcal{T}_{b_1 b_2 b_3 b_4} \cap \ldots,$$

defines a unique point which, by construction, lies on $\mathcal{C}$:

$$f(0_{\dot{2}} b_1 b_2 b_3 \ldots) = \mathcal{T}_{b_1} \cap \mathcal{T}_{b_1 b_2} \cap \mathcal{T}_{b_1 b_2 b_3} \cap \mathcal{T}_{b_1 b_2 b_3 b_4} \cap \ldots .$$

f is continuous: If $|t_1 - t_2| < 1/2^n$, then, $f(t_1), f(t_2)$ lie, at worst, in two adjacent triangles $\mathcal{T}_{b_1 b_2 b_3 \ldots b_n}$ and their distance cannot exceed the sum of the lengths of the two longest sides. Since the sides shrink to zero, continuity readily follows. □

f is injective: Any point on $\mathcal{C}$ has to lie in $\mathcal{T}_0$ or $\mathcal{T}_1$. In turn, it has to lie in one of the four triangles $\mathcal{T}_{b_1 b_2}$, and, in turn, in one of the eight triangles $\mathcal{T}_{b_1 b_2 b_3}$, etc.... (Note that it cannot lie, at the same time, in two triangles of the same partition, except when it is the one point that is common to two adjacent triangles, in which case it has, as the unique preimage in [0,1], the point that is common to two adjacent closed subintervals.) With this sequence of nested closed triangles there corresponds a unique nested sequence of closed intervals $[b_1/2, b_1/2 + 1/2] \supset [b_1/2 + b_2/4, b_1/2 + b_2/4 + 1/4] \supset [b_1/2 + b_2/4 + b_3/8, b_1/2 + b_2/4 + b_3/8 + 1/8] \supset \ldots$ which defines a unique point in $[0,1]$. □

We summarize our result in the following:

(8.3) Theorem. *The curve $\mathcal{C}$ that is obtained by the process that is outlined for the first two steps in Fig. 8.3.2 is a Jordan curve with a positive two-dimensional Lebesgue measure as given in (8.3.2).*

Knopp's objective was not only to find a new Osgood curve but to obtain a representation of nowhere differentiable curves with two-dimensional Lebesgue measure 0 at the one extreme, two-dimensional Lebesgue measure 1 at the other extreme, and all the numbers in between.

At the one extreme, we start out with an isosceles triangle that has an angle of 120° at the top, take $r_j = 1/3$, and remove isosceles triangles at every step. This will make all the remaining triangles similar to the initial triangles, and we obtain a Jordan curve of two-dimensional Lebesgue measure

$$\Lambda_2(\mathcal{C}) = \prod_{j=1}^{\infty}(1 - 1/3) = \lim_{n\to\infty}(2/3)^n = 0.$$

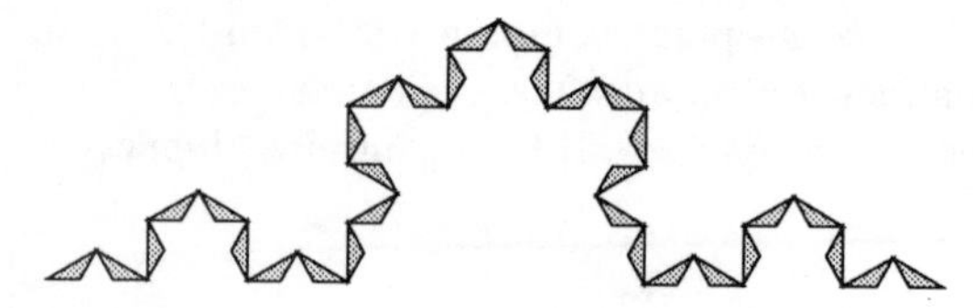

Fig. 8.3.3. Fourth Step in Knopp's Generation of the von Koch Curve

We have depicted the fourth step in the generation of this curve in Fig. 8.3.3. The limiting curve is the *von Koch curve* which was already mentioned in Section 5.4 and will be discussed in greater detail in Section 9.1. (See also Fig. 9.1.2.)

At the other extreme, we start out with an isosceles right triangle with a base of length 2 and set $r_j = r^2/j^2$. We have, from (8.3.2),

$$\Lambda_2(\mathcal{C}) = \frac{\sin(\pi r)}{\pi r} \longrightarrow 1 \text{ as } r \longrightarrow 0,$$

i.e., in the limit, we obtain a space-filling curve. Comparing Fig. 8.3.2 with Fig. 4.2.1 reveals the limiting curve as the Sierpiński-Knopp curve of Chapter 4. Incidentally, it was this construction that led Knopp to the complex representation (4.2.4) of the Sierpiński-Knopp curve.

Recently, K. Stromberg and his student, S. Tseng, observed that the examples of Sierpiński and Knopp, while having the property that every part of their curves is again an Osgood curve, do not guarantee another desirable feature: a homogeneity property in the following sense: For $\beta \in (0,1)$, they ask that for every subset $\mathcal{E} \subseteq [0,1], \Lambda_2(f(\mathcal{E})) = \beta\Lambda_1(\mathcal{E})$, whereupon they proceed to construct an Osgood curve with this very property (Stromberg and Tseng [1]).

8.4. Other Osgood Curves

In the preceding sections, we discussed families of Osgood curves, one of which had the Peano curve as a limiting arc, while the other had the Sierpińki-Knopp curve as a limiting arc. In Fig. 8.4.1 we indicate for the first two steps, how a family of Osgood curves with the Hilbert-curve as a limiting arc may be obtained (Gelbaum and Olmsted [1], p. 137). Observe that this construction requires the use of "joins."

Gelbaum and Olmsted ([1], p. 135) present another (very complicated) example, modifying the generating process for the Hilbert curve by cutting out "channels" between adjacent squares that do not correspond to adjacent subintervals "to remove irrelevant adjacencies." Their construction does not use "joins."

T. Lance and E. Thomas [1], reverting to the use of "joins," propose a construction (Lance and Thomas [1]) which we modified (Sagan [10]) to

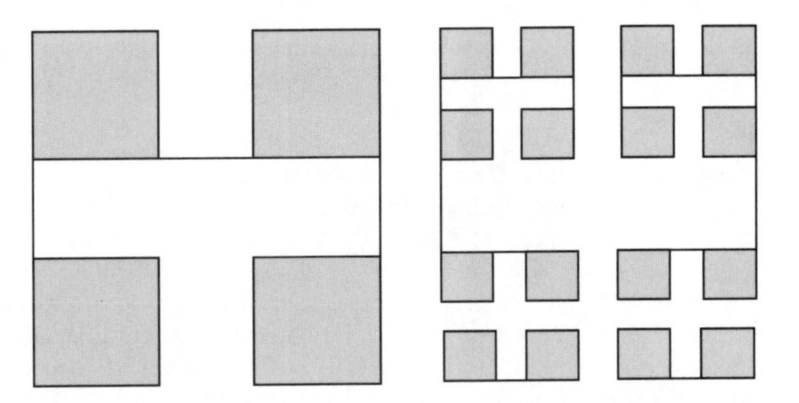

Fig. 8.4.1. Generating an Osgood Curve with "Joins" with the Hilbert Curve as a Limiting Arc

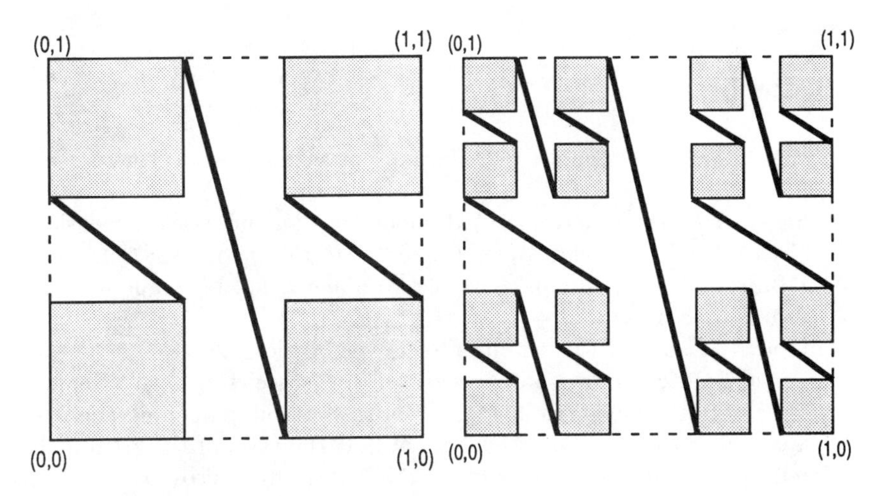

Fig. 8.4.2. Generating a Family of Osgood Curves with the Lebesgue Curve as a Limiting Arc

obtain a family of Osgood curves with the Lebesgue curve as a limiting arc. In Fig. 8.4.2, we indicated the first two steps in the generating process. We choose the width of the bars of the first cross-shaped region to be $(1-\sqrt{\lambda})/3$, the width of the next one $(1-\sqrt{\lambda})/9$, then $(1-\sqrt{\lambda})/27$, etc.... As before, we denote the region consisting of the remaining squares and joins at the nth step by $\mathcal{A}_n(\lambda)$ and have $J_2(\mathcal{A}_n(\lambda)) = \{1-[1-(2/3)^n][1-\sqrt{\lambda}]\}^2$. Hence, we obtain for the two-dimensional Lebesgue measure of $\mathcal{C}(\lambda) = \bigcap_{n=1}^{\infty} \mathcal{A}_n(\lambda)$:

$$\Lambda_2(\mathcal{C}(\lambda)) = \lim_{n\to\infty} [\sqrt{\lambda} + (2/3)^n(1-\sqrt{\lambda})]^2 = \lambda.$$

We parametrize $\mathcal{C}(\lambda)$ as we parametrized Osgood's curve in Section 8.2 but instead of Γ_8 we now use the Cantor set Γ. We map the intervals that are removed in the construction of the Cantor set linearly onto the "joins" and the remaining closed intervals into the remaining squares, and we see, as in Section 8.2, that the mapping is continuous and injective.

Summarizing, we can state:

(8.4) Theorem. *The generating process that is outlined for the first two steps in Fig. 8.4.2 leads, for any $\lambda \in (0,1)$, to an Osgood curve of two-dimensional Lebesgue measure λ.*

If we choose $\lambda = 0$, i.e., remove cross-shaped regions having bars of widths $1/3, 1/9, 1/27, \ldots$, we obtain a Jordan curve of two-dimensional Lebesgue measure zero. (In this case, $\mathcal{C}(0)$ is the union of the cartesian product of the Cantor set with itself and the appropriate joins). If we let $\lambda \longrightarrow 1$, the Lebesgue curve of Chapter 5 is obtained as the limiting arc (see Section 5.5 and, in particular, Fig. 5.5.1, 5.5.2, and 5.5.3).

8.5. Problems

1. Prove that there cannot be a Jordan curve that is everywhere dense in $\mathcal{Q}$.
2. Show: If, in the generation of Osgood's Jordan curve, the dimensions of the remaining squares (see Fig. 8.2) are taken to be $18^{-(n/2)}, n = 1, 2, 3, \ldots$, then a Jordan curve of two-dimensional Lebesgue measure zero is obtained.
3. Find $\Lambda_1(\Gamma_8)$. (For the definition of Γ_8 see Section 8.2.)
4. Let γ denote the top obtuse angle in an isosceles triangle. Remove an isosceles triangle (as in Fig. 8.3.1) so that the remaining triangles are similar to the original one. Show that the ratio of the area of the removed triangle to the area of the original triangle is $\cos\gamma/(\cos\gamma - 1)$. (Note that this ratio is positive, since we have $\pi/2 < \gamma < \pi$.)
5. Continue the process outlined in Problem 4 ad infinitum, and show that a "von Koch type" curve of two-dimensional Lebesgue measure zero is obtained in the limit.
6. Construct a closed Jordan curve as follows: Divide a circular ring by radii into an even number $2m$ of congruent sections. Map $\mathcal{Q}$, carrying Osgood's Jordan curve $\mathcal{C}(\lambda)$ of two-dimensional Lebesgue measure $\lambda > 0$ of Section 8.2, conformally onto one of these sections with corners going into corners. Let $\mathcal{C}'(\lambda)$ denote the image of $\mathcal{C}(\lambda)$. Reflect $\mathcal{C}'(\lambda)$ on the boundary with the adjacent section to obtain $\mathcal{C}''(\lambda)$, and reflect again and again until, after $2m$ steps, $\mathcal{C}^{(2m)} = \mathcal{C}(\lambda)$. Note that the endpoint of $\mathcal{C}^{(2m-1)}(\lambda)$ coincides with the beginning point of $\mathcal{C}(\lambda)$. In this manner, a closed Jordan curve has been constructed. By Jordan's curve theorem, it partitions the plane into an interior and an exterior. Show that the inner measure of the interior (which is simply connected) is less than the outer measure of the interior.

7. Let $\mathcal{C}(\lambda)$ denote the Osgood curve of Section 8.2, and let $\Gamma_2(\lambda)$ denote the Cantor-type set that is obtained by first removing from $\mathcal{I}$ two open intervals of length $(1-\sqrt{\lambda})/4$ each, leaving three closed intervals of equal length, and then removing from each of these closed intervals two open intervals of length $(1-\sqrt{\lambda})/24$ each, and repeating the process with $(1-\sqrt{\lambda})/144$, etc.... Show that $\Lambda_2(\mathcal{C}(\lambda)) = \Lambda_2(\Gamma_2(\lambda) \times \Gamma_2(\lambda))$.
8. Same as in Problem 7, using the Osgood curve of Section 8.4, the generation of which was illustrated in Fig. 8.4.2 and a Cantor-type set $\Gamma_1(\lambda)$, which is obtained from $\mathcal{I}$ by first removing an open interval of length $(1-\sqrt{\lambda})/3$, then from each of the remaining two congruent closed intervals an open interval of length $(1-\sqrt{\lambda})/9$, then $(1-\sqrt{\lambda})/27$, etc.
9. Show: when in the construction of Knopp's Osgood curve of Section 8.3, one starts out with a triangle that has an obtuse angle γ at the top, with a base c and a smallest angle α, and keeps removing triangles so that the oblique triangles with the oblique angle not exceeding γ remain, then none of the angles in all the remaining triangles can be less than α, and none of the bases at the nth step can exceed $c\cos^n\alpha$.
10. Construct a family of three-dimensional Osgood curves that has the three-dimensional Hilbert curve of Section 2.8 as a limiting arc.

Chapter 9

Fractals

9.1. Examples

If we apply the similarity transformations

$$
\begin{aligned}
\xi' &= \frac{1}{3}\xi \\
\xi' &= \frac{1}{3}(\xi + 2)
\end{aligned}
\tag{9.1.1}
$$

to the interval $\mathcal{I}$ in Fig. 9.1.1(a), we obtain the configuration in Fig. 9.1.1(b). If we apply (9.1.1) to the configuration in Fig. 9.1.1(b), we obtain the configuration in Fig. 9.1.1(c). Applying it to Fig. 9.1.1(c), we obtain the configuration in Fig. 9.1.1(d), etc. If we carry this on ad infinitum, we arrive at the Cantor set of Section 5.1 (or *Cantor dust*, as B. Mandelbrot so aptly called it).

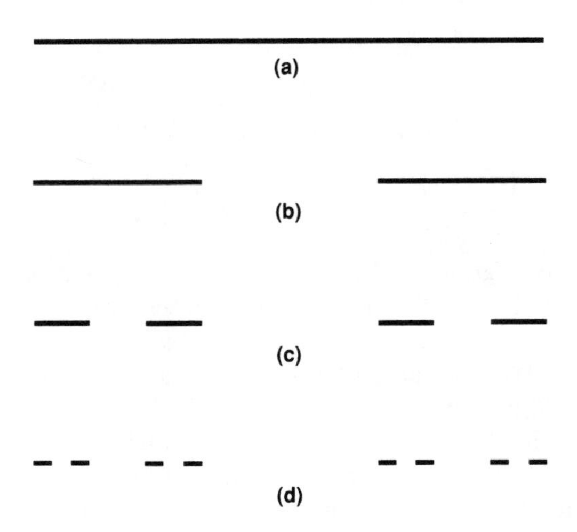

Fig. 9.1.1. Generating the Cantor Set

Next, we consider the four similarity transformations

$$\begin{aligned}
\begin{pmatrix} \xi' \\ \eta' \end{pmatrix} &= \frac{1}{3}\begin{pmatrix} 1 & 0 \\ 0 & 1 \end{pmatrix}\begin{pmatrix} \xi \\ \eta \end{pmatrix} \\
\begin{pmatrix} \xi' \\ \eta' \end{pmatrix} &= \frac{1}{3}\begin{pmatrix} 1/2 & -\sqrt{3}/2 \\ \sqrt{3}/2 & 1/2 \end{pmatrix}\begin{pmatrix} \xi \\ \eta \end{pmatrix} + \frac{1}{3}\begin{pmatrix} 1 \\ 0 \end{pmatrix} \\
\begin{pmatrix} \xi' \\ \eta' \end{pmatrix} &= \frac{1}{3}\begin{pmatrix} 1/2 & \sqrt{3}/2 \\ -\sqrt{3}/2 & 1/2 \end{pmatrix}\begin{pmatrix} \xi \\ \eta \end{pmatrix} + \frac{1}{3}\begin{pmatrix} 3/2 \\ \sqrt{3}/2 \end{pmatrix} \\
\begin{pmatrix} \xi' \\ \eta' \end{pmatrix} &= \frac{1}{3}\begin{pmatrix} 1 & 0 \\ 0 & 1 \end{pmatrix}\begin{pmatrix} \xi \\ \eta \end{pmatrix} + \frac{1}{3}\begin{pmatrix} 2 \\ 0 \end{pmatrix}
\end{aligned} \tag{9.1.2}$$

and apply them again to the interval $\mathcal{I}$. After one application, we obtain the configuration in Fig. 9.1.2(a), after application to Fig. 9.1.2(a), the one in Fig. 9.1.2(b). We keep applying (9.1.2) to the preceding result to obtain Fig. 9.1.2(c), Fig. 9.1.2(d), etc.... After infinitely many steps, we wind up with the von Koch curve, which Richard F. Voss called one of the "early mathematical monsters." (Peitgen and Saupe [1], p. 26; see also Section 8.3.) Voss adds that "although the algorithm for generating the von Koch curve is concise, simple to explain, and easily computed, there is no algebraic formula that specifies the points of the curve." (loc. cit. pp. 27–28). But there is: E. Cesàro and K. Knopp derived such a formula (Cesàro [2], Knopp [1]. See also Sagan [12].)

In the preceding two cases we used $\mathcal{I}$ as the initial set (or *Leitmotiv*, as we called it in Section 2.6). In the next example, we use an equilateral triangle of sidelength 1 as initial set [Fig. 9.1.3(a)] and subject it to the following three transformations:

$$\begin{aligned}
\begin{pmatrix} \xi' \\ \eta' \end{pmatrix} &= \frac{1}{2}\begin{pmatrix} 1 & 0 \\ 0 & 1 \end{pmatrix}\begin{pmatrix} \xi \\ \eta \end{pmatrix} \\
\begin{pmatrix} \xi' \\ \eta' \end{pmatrix} &= \frac{1}{2}\begin{pmatrix} 1 & 0 \\ 0 & 1 \end{pmatrix}\begin{pmatrix} \xi \\ \eta \end{pmatrix} + \frac{1}{2}\begin{pmatrix} 1 \\ 0 \end{pmatrix} \\
\begin{pmatrix} \xi' \\ \eta' \end{pmatrix} &= \frac{1}{2}\begin{pmatrix} 1 & 0 \\ 0 & 1 \end{pmatrix}\begin{pmatrix} \xi \\ \eta \end{pmatrix} + \frac{1}{2}\begin{pmatrix} 1/2 \\ 1 \end{pmatrix}
\end{aligned} \tag{9.1.3}$$

to obtain the configuration in Fig. 9.1.3(b). Application of (9.1.3) to Fig. 9.1.3(b) yields Fig. 9.1.3(c). After infinitely many steps we obtain the *Sierpiński triangle*, or the *Sierpiński gasket*, as it is also called (Sierpiński [5]). We see that, at each step, the remaining triangles are dissected into four congruent triangles and the one in the middle is eliminated. In the fractal literature, the eliminated portions are often referred to as the *trema* (which, apparently, is derived from the Greek $\tau\rho\upsilon\mu\alpha\lambda\acute{\iota}\alpha$ = hole). In our first example of the Cantor set, the middle thirds are the tremas.

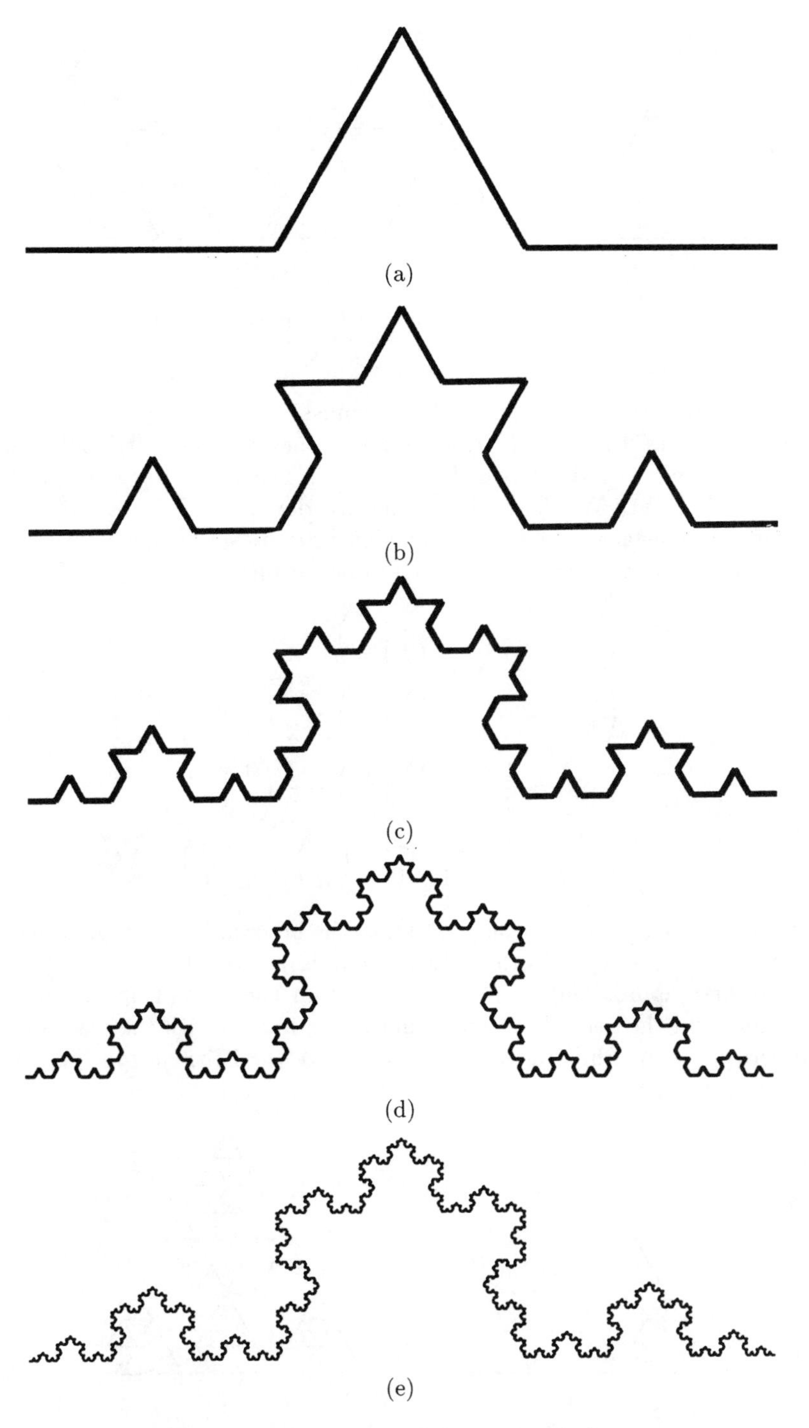

Fig. 9.1.2. Generation of the von Koch Curve

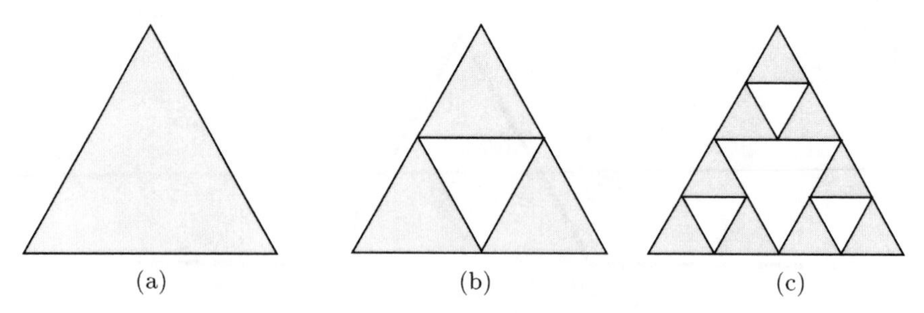

Fig. 9.1.3. Generation of the Sierpiński Gasket

An interesting variation of the Sierpiński triangle may be obtained, if one dissects the initial triangle into nine congruent triangles and removes three, as indicated in Fig. 9.1.4(a). The next step is illustrated in Fig. 9.5.1(b). We leave it to the reader to write the similarity transformations that generate this *modified Sierpiński triangle.*

If we apply the four similarity transformations

$$\begin{aligned}
\begin{pmatrix}\xi'\\ \eta'\end{pmatrix} &= \frac{1}{2}\begin{pmatrix}0 & 1\\ 1 & 0\end{pmatrix}\begin{pmatrix}\xi\\ \eta\end{pmatrix}\\
\begin{pmatrix}\xi'\\ \eta'\end{pmatrix} &= \frac{1}{2}\begin{pmatrix}1 & 0\\ 0 & 1\end{pmatrix}\begin{pmatrix}\xi\\ \eta\end{pmatrix}+\frac{1}{2}\begin{pmatrix}0\\ 1\end{pmatrix}\\
\begin{pmatrix}\xi'\\ \eta'\end{pmatrix} &= \frac{1}{2}\begin{pmatrix}1 & 0\\ 0 & 1\end{pmatrix}\begin{pmatrix}\xi\\ \eta\end{pmatrix}+\frac{1}{2}\begin{pmatrix}1\\ 1\end{pmatrix}\\
\begin{pmatrix}\xi'\\ \eta'\end{pmatrix} &= \frac{1}{2}\begin{pmatrix}0 & -1\\ -1 & 0\end{pmatrix}\begin{pmatrix}\xi\\ \eta\end{pmatrix}+\frac{1}{2}\begin{pmatrix}2\\ 1\end{pmatrix}
\end{aligned} \tag{9.1.4}$$

(which are the transformations (2.4.1)) to the interval $\mathcal{I}$, we obtain the first approximating polygon to the Hilbert curve in Fig. 2.6.1. We apply (9.1.4) to the first approximating polygon to obtain the second approximating polygon, and then the third, etc... until we wind up with the trace of the Hilbert curve, which is the square $\mathcal{Q}$. If we use the Leitmotiv in Fig. 2.6.2,

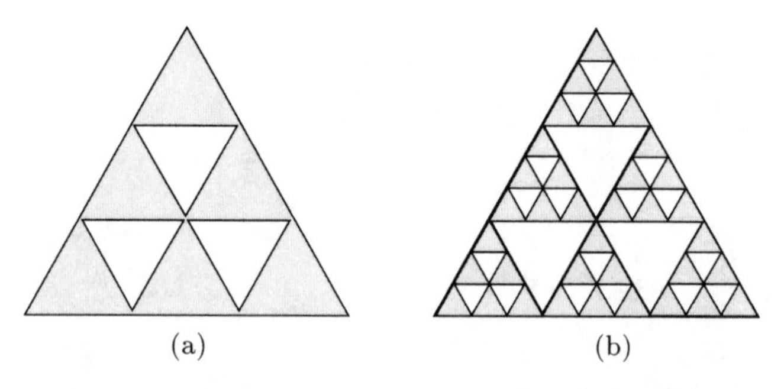

Fig. 9.1.4. Generation of the Modified Sierpiński Triangle

we still wind up with $\mathcal{Q}$, and the same happens if we start out with the Leitmotiv in Fig. 2.6.3—or any other Leitmotiv—as long as it is a non-empty compact subset of $\mathbb{E}^2$, as we will see in Section 9.3.

We can do the same with the similarity transformations (3.4.1) to obtain $\mathcal{Q}$ as the trace of the Peano curve and with (4.3.1) to obtain $\mathcal{T}$ as the trace of the Sierpiński-Knopp curve.

The limit sets that are obtained in the preceding cases are special cases of what are called *self-similar fractals*. More generally, a self-similar fractal is obtained by applying a number of similarity transformations

$$x' = r_i U_i x + b_i, \quad i = 1, 2, 3, \ldots, n \tag{9.1.5}$$

(where $x, x', b_i \in \mathbb{E}^m, 0 < r_i < 1$, and where the U_i represent orthogonal $m \times m$ matrices) to a compact, non-empty subset of $\mathbb{E}^m$, and then again to the union of the images that were obtained at the first step, then again and again, ad infinitum. The sequence of sets thus generated converges in a certain sense to a unique, compact, non-empty subset of $\mathbb{E}^m$, such as the Cantor set in our first example, the von Koch curve in our second example, the Sierpiński triangle in our third example, etc.... .

In Section 9.2 we will explain what we mean by "convergence" of a sequence of compact subsets of $\mathbb{E}^m$ (or any complete metric space, for that matter), and in Section 9.3 we will establish the existence of a unique limit set.

There is a vast, superbly illustrated, literature on fractals available. We refer the reader to Barnsley [1], Edgar [1], Mandelbrot [1], Peitgen and Richter [1], and Peitgen and Saupe [1]. A brief history of fractals, where their origin is traced to H. Poincaré, R.R. Fricke, and Felix Klein, may be found in Peitgen and Saupe [1]. Because of the nature of their generation by iteration, the sets that are obtained after each application of the similarity transformations (9.1.5) lend themselves quite readily to generation by computer. Computer programs may be found, for example, in Barnsley [1], Edgar [1], and Peitgen and Saupe [1].

To give credit where it is due, let us point out that W. Wunderlich, in a 1954 paper (Wunderlich [1]) that apparently attracted little attention, has already used this process of similarity transformations to generate the von Koch curve, the Sierpiński gasket, the Hilbert curve, and the Sierpiński-Knopp curve. In the same paper, he used affine transformations to generate the Steinitz curves (Steinitz [1]), the Bolzano-Kowalewski curve (Kowalewski [1]), and the Rham curve (Rham [1]).

9.2. The Space where Fractals are Made

With the abbreviating notation

$$r_i U_i x + b_i \stackrel{\triangle}{=} \mathfrak{F}_i(x),$$

we may write the system of similarity transformations (9.1.5) as

$$x' = \mathfrak{F}_i(x), \quad i = 1, 2, 3, \ldots, n. \tag{9.2.1}$$

M. Barnsley introduced the name *iterated function system* for such a set of similarity transformations, and we will follow this custom.

Since orthogonal transformations are length-preserving, we have

$$||\mathfrak{F}_i(x_1) - \mathfrak{F}_i(x_2)|| = r_i||U_i x_1 - U_i x_2|| = r_i||x_1 - x_2||,$$

which explains why r_i is called the *reduction ratio of* $\mathfrak{F}_i$.

We apply the iterated function system (9.2.1) to some non-empty compact subset $\mathcal{A}_0 \subset \mathbb{E}^m$ and then again to

$$\mathcal{A}_1 = \bigcup_{j=1}^{n} \mathfrak{F}_{j*}(\mathcal{A}_0),$$

which as a finite union of continuous images of compact sets is, again, compact. We repeat the process again and again, in order to generate an infinite sequence of compact sets

$$\mathcal{A}_{k+1} = \bigcup_{j=1}^{n} \mathfrak{F}_{j*}(\mathcal{A}_k), \quad k = 0, 1, 2, 3, \ldots .$$

We will show in the next section that this sequence converges in a sense, that will be specified below, to a unique limit set, the so-called *invariant attractor set*, and that this limit set is independent of the choice of the initial set.

Since we are now working in a space, the elements of which are non-empty compact subsets of $\mathbb{E}^m$, we have to define what we mean by convergence. This, in turn, requires an understanding of what we mean by the "distance" of two non-empty compact subsets of $\mathbb{E}^m$. What we are about to develop is not only valid for euclidean spaces but, quite generally, for complete metric spaces S (every Cauchy sequence in S converges to an element in S). We will henceforth use the notation S, but the reader may think of it as an m-dimensional euclidean space.

First, let us introduce the notation $K(S)$ for the set of all non-empty compact subsets of S. We will now define the *Hausdorff distance* between two sets $\mathcal{A}, \mathcal{B} \subseteq K(S)$ in terms of the concept of *open δ-neighborhoods of compact subsets of S*:

(9.2.1) Definition. *Let $\rho(x, y)$ denote the distance from x to y in S and let $\mathcal{A}$ denote a non-empty compact subset of S. We say that $x \in N_\delta(\mathcal{A})$ (δ-neighborhood of $\mathcal{A}$) if there is a $y \in \mathcal{A}$ such that $\rho(x, y) < \delta$.*

For example, $\{x \in S|\rho(0, x) < 1 + \varepsilon\}$ is an ε-neighborhood of the unit ball $\{x\varepsilon S|\rho(0, x) < 1\}$. A neighborhood of a set is sometimes referred to as a *dilation* of the set.

(9.2.2) Definition. *The Hausdorff distance between two non-empty compact sets $\mathcal{A}, \mathcal{B} \subset K(S)$ is defined by*

$$d(\mathcal{A}, \mathcal{B}) = \inf\{\delta | \mathcal{A} \subseteq N_\delta(\mathcal{B}) \ \&\ \mathcal{B} \subseteq N_\delta(\mathcal{A})\}.$$

This is, indeed, a distance:

(9.2.1) Lemma. *The Hausdorff distance, as defined in Definition 9.2.2, satisfies the distance postulates.*

Proof. (1) By definition, $d(\mathcal{A}, \mathcal{B}) \geq 0$. $d(\mathcal{A}, \mathcal{B}) = 0$ if and only if $\mathcal{A} = \mathcal{B}$: If $\mathcal{A} = B$, then, for any $\varepsilon > 0$, $\mathcal{A} \subseteq N_\varepsilon(\mathcal{B})$ and $\mathcal{B} \subseteq N_\varepsilon(A)$. Hence, $d(\mathcal{A}, \mathcal{B}) = 0$. Conversely, suppose that $d(\mathcal{A}, \mathcal{B}) = 0$. Whenever $x \in \mathcal{A}$, we have for every $\varepsilon > 0$ that $x \in N_\varepsilon(\mathcal{B})$, i.e., $\rho(x, y) < \varepsilon$ for some $y \in \mathcal{B}$. Since $\mathcal{B}$ is compact, $x \in \mathcal{B}$. Hence, $\mathcal{A} \subseteq \mathcal{B}$. By symmetric reasoning, $\mathcal{B} \subseteq \mathcal{A}$ and we have $\mathcal{A} = \mathcal{B}$.

(2) It follows directly from the definition that $d(\mathcal{A}, \mathcal{B}) = d(\mathcal{B}, \mathcal{A})$.

(3) The triangle inequality is satisfied: Let $\mathcal{A}, \mathcal{B}, \mathcal{C}$ denote non-empty compact subsets of S. Let $x \in \mathcal{A}$. For arbitrary $\varepsilon > 0$, there is a $y \in \mathcal{B}$ such that $\rho(x, y) < d(\mathcal{A}, \mathcal{B}) + \varepsilon$ and a $z \in \mathcal{C}$ such that $\rho(y, z) < d(\mathcal{B}, \mathcal{C}) + \varepsilon$. Since $\rho(x, z) \leq \rho(x, y) + \rho(y, z) < d(\mathcal{A}, \mathcal{B}) + d(\mathcal{B}, \mathcal{C}) + 2\varepsilon$, we have $\mathcal{A} \subset N_{d(\mathcal{A},\mathcal{B})+d(\mathcal{B},\mathcal{C})+2\varepsilon}(\mathcal{C})$ and, by the same token, $\mathcal{C} \subset N_{d(\mathcal{A},\mathcal{B})+d(\mathcal{B},\mathcal{C})+2\varepsilon}(\mathcal{A})$. Hence, by definition (9.2.2), $d(\mathcal{A}, \mathcal{C}) \leq d(\mathcal{A}, \mathcal{B}) + d(\mathcal{B}, \mathcal{C}) + 2\varepsilon$. This is true for all $\varepsilon > 0$ and, consequently, $d(\mathcal{A}, \mathcal{C}) \leq d(\mathcal{A}, \mathcal{B}) + d(\mathcal{B}, \mathcal{C})$. □

We obtain directly from the definition of the Hausdorff distance in Definition 9.2.2 that

(9.2.2) Lemma. $d(\mathcal{A}, \mathcal{B}) < \varepsilon$ *if and only if* $\mathcal{A} \subseteq N_\varepsilon(\mathcal{B})$ *and* $\mathcal{B} \subseteq N_\varepsilon(\mathcal{A})$.

In order to establish the existence and uniqueness of the attractor set, we will demonstrate that $\mathfrak{F}_* : K(S) \longrightarrow K(S)$, defined by $\mathfrak{F}_*(\mathcal{A}) = \bigcup_{i=1}^n \mathfrak{F}_{i*}(\mathcal{A})$, is a contraction mapping and then use the contraction mapping theorem. Application of the contraction mapping theorem, however, requires the completeness of the space $K(S)$. The objective of this section is to demonstrate this completeness. The following two lemmas will prove helpful in this endeavor.

(9.2.3) Lemma. *If $\mathcal{A}, \mathcal{B} \subseteq K(S)$ and $d(\mathcal{A}, \mathcal{B}) < \varepsilon$, then, for any fixed $a \in \mathcal{A}$, there is a $b \in \mathcal{B}$ such that $\rho(a, b) < \varepsilon$.*

Proof. By hypothesis, $\inf\{\delta | \mathcal{A} \subseteq N_\delta(\mathcal{B}) \& \mathcal{B} \subseteq N_\delta(\mathcal{A})\} < \varepsilon$. Hence, $a \in \mathcal{A} \subseteq N_\varepsilon(\mathcal{B})$, meaning that there is some $b \in \mathcal{B}$ such that $\rho(a, b) < \varepsilon$. □

In order to show that $K(S)$ is complete, we have to demonstrate that every Cauchy sequence $\{\mathcal{A}_n\}$ in $K(S)$ converges to a non-empty, compact subset of S. It stands to reason that the limit set will be made up of the limits of all Cauchy sequences of points from S that can be formed by picking the first element from $\mathcal{A}_1$, the second element from $\mathcal{A}_2$, etc... :

$$\mathcal{A} = \{x \in S | x \text{ is the limit of a sequence } \{x_n\}, x_n \in \mathcal{A}_n, n = 1, 2, 3, \ldots\}.$$

In order to show that this set is not empty, we need:

(9.2.4) Lemma. *If $\{\mathcal{A}_n\}$ is a Cauchy sequence in the metric space $K(S)$, with the metric defined by the Hausdorff distance, then there exists a sequence $\{x_n\} \to x \in S, x_n \in \mathcal{A}_n$.*

Proof. Since $\{\mathcal{A}_n\}$ is a Cauchy sequence, we have for every $\varepsilon > 0$ an N_ε such that $d(\mathcal{A}_n, \mathcal{A}_m) < \varepsilon$ for all $m, n > N_\varepsilon$. Hence, we can pick an increasing sequence $N_1 < N_2 < N_3 < \cdots < N_j < \cdots$ such that

$$d(\mathcal{A}_n, \mathcal{A}_{N_j}) < \varepsilon/2^j \quad \text{for all} \quad n > N_j, j = 1, 2, 3, \ldots .$$

Pick an arbitrary $x_{N_1} \in \mathcal{A}_{N_1}$. By construction, $d(\mathcal{A}_{N_1}, \mathcal{A}_{N_2}) < \varepsilon/2$. By Lemma 9.2.3, there is an $x_{N_2} \in \mathcal{A}_{N_2}$ such that $\rho(x_{N_1}, x_{N_2}) < \varepsilon/2$. Again, by Lemma 9.2.3, there is an $x_{N_3} \in \mathcal{A}_{N_3}$ such that $\rho(x_{N_2}, x_{N_3}) < \varepsilon/4$, etc.... Eventually, there is an $x_{N_j} \in \mathcal{A}_{N_j}$ such that $\rho(x_{N_{j-1}}, x_{N_j}) < \varepsilon/2^{j-1}$, etc.... $\{x_{N_j}\}$ is a Cauchy sequence:

$$\begin{aligned}\rho(x_{N_m}, x_{N_n}) &\le \rho(x_{N_m}, x_{N_{m+1}}) + \cdots + \rho(x_{N_{n-1}}, x_{N_n}) \\ &\le \varepsilon/2^m + \cdots + \varepsilon/2^{n-1} < \varepsilon/2^{m-1}.\end{aligned}$$

We now proceed to fill out this sequence to obtain one, as required. For $n = 1, 2, 3, \ldots, N_1$, we pick $\bar{x}_n \in \mathcal{A}_n$ arbitrarily. Let $\bar{x}_{N_j} = x_{N_j}$ for $j = 2, 3, 4, \ldots$. We pick an $\bar{x}_n \in \mathcal{A}_n$ for all n for which $N_{N_1} < n < N_{N_2}$, such that $\rho(\bar{x}_{N_1}, \bar{x}_n) < \varepsilon/2$. This can be done in view of Lemma 9.2.3. We continue in this manner: Eventually, we pick for all n for which $N_j < n < N_{j+1}$, an $\bar{x}_n \in \mathcal{A}_n$ such that $\rho(\bar{x}_{N_j}, \bar{x}_n) < \varepsilon/2^j$, etc.... The sequence $\{\bar{x}_n\}$ coincides with the sequence $\{x_{N_j}\}$ for $n = N_1, N_2, N_3, \ldots$ and $\bar{x}_n \in \mathcal{A}_n$. It is a Cauchy sequence by construction and, since S is complete, $\lim_{n\to\infty} \bar{x}_n = x (\in S)$ exists. □

Corollary to Lemma 9.2.4. *If $\{\mathcal{A}_n\}$ is a Cauchy sequence in $K(S)$, then*

$$\mathcal{A} = \{x \in S | x \textit{ is the limit of a sequence } \{x_n\},\ x_n \in \mathcal{A}_n,\ n = 1, 2, 3, \ldots\}$$

is not empty.

We are now ready to establish the main result of this section. We traced it back to W. Blaschke—"Selection Theorem of Blaschke"—who first mentioned that one can select from an infinite set of uniformly bounded convex bodies a sequence that converges to a convex body. (Blaschke [1], p. 200. A proof is to be found in Blaschke [2], p. 62.)

(9.2) Theorem. *The space $K(S)$ of compact non-empty subsets of S with the Hausdorff metric is complete, i.e., every Cauchy sequence $\{\mathcal{A}_n\}$ with $\mathcal{A}_n \in K(S)$ converges to some $\mathcal{A} \in K(S)$.*

Proof. We will show that $\{\mathcal{A}_n\} \to \mathcal{A}$, where $\mathcal{A}$ is defined in the Corollary to Lemma 9.2.4, and that $\mathcal{A}$ is compact.

To show that $\{\mathcal{A}_n\} \to \mathcal{A}$, we will demonstrate that for any $\varepsilon > 0$, there is an N_ε such that $d(\mathcal{A}_n, \mathcal{A}) < \varepsilon$ as long as $n > N_\varepsilon$, where $\mathcal{A} = \{x \in S | x \text{ is the limit of a sequence } \{x_n\}, x_n \in \mathcal{A}_n, n = 1, 2, 3, \ldots\}$. By Lemma 9.2.2, this is equivalent to $\mathcal{A} \subseteq N_\varepsilon(\mathcal{A}_n)$ and $\mathcal{A}_n \subseteq N_\varepsilon(\mathcal{A})$ for all $n > N_\varepsilon$.

(a) To show that $\mathcal{A} \subseteq N_\varepsilon(\mathcal{A}_n)$, we proceed as follows: For $x \in \mathcal{A}$, there is a sequence $\{x_n\} \to x$ with $x_n \in \mathcal{A}_n$. Pick N so that $\rho(x_n, x) < \varepsilon/2$ and $d(\mathcal{A}_n, \mathcal{A}_m) < \varepsilon/2$ for all $n, m > N$. If $x_n \in \mathcal{A}_n$, there is, by Lemma 9.3.3, an $x_m \in \mathcal{A}_m$ such that $\rho(x_n, x_m) < \varepsilon/2$. Hence, $\rho(x_m, x) \le \rho(x_m, x_n) + \rho(x_n, x) < \varepsilon$. Therefore, $x \in N_\varepsilon(\mathcal{A}_m)$. Hence, $\mathcal{A} \subseteq N_\varepsilon(\mathcal{A}_m)$.

(b) We clinch the argument by showing that $\mathcal{A}_n \subseteq N_\varepsilon(\mathcal{A})$ for sufficiently large n. Choose a sequence $N_1 < N_2 < N_3 < \cdots < N_j < \cdots$ such that $d(\mathcal{A}_{N_j}, \mathcal{A}_m) < \varepsilon/2^{j+1}$ for all $m > N_j, j = 1, 2, 3, \ldots$. Let $y \in \mathcal{A}_{N_1}$ and let $n = N_1$. By Lemma 9.2.3, we can pick $x_{N_2} \in \mathcal{A}_{N_2}$, such that $\rho(y, x_{N_2}) < \varepsilon/2^2, x_{N_3} \in \mathcal{A}_{N_3}$ such that $\rho(x_{N_2}, x_{N_3}) < \varepsilon/2^3$, etc.... In general, we pick $x_{N_{j+1}} \in \mathcal{A}_{N_{j+1}}$ such that $\rho(x_{N_j}, x_{N_{j+1}}) < \varepsilon/2^{j+1}$. $\{x_{N_j}\}$ is a Cauchy sequence and we can fill it up, as in the proof of Lemma 9.2.4, to obtain a Cauchy sequence $\{x_k\} \to x \in \mathcal{A}$, which coincides with $\{x_{N_j}\}$ for $k = N_j$ and $x_n \in \mathcal{A}_n$. Hence, $\{x_{N_j}\} \to x$. By construction,

$$\begin{aligned}\rho(y, x_{N_j}) &\le \rho(x_{N_1}, x_{N_2}) + \rho(x_{N_2}, x_{N_3}) + \cdots + \rho(x_{N_{j-1}}, x_{N_j}) \\ &< (\varepsilon/4)(1 + 1/2 + 1/4 + \cdots) = \varepsilon/2.\end{aligned}$$

Hence,

$$\rho(y, x) = \lim_{j \to \infty} \rho(y, x_{N_j}) \le \varepsilon.$$

This is true for any $y \in \mathcal{A}_n$ and we have $\mathcal{A}_n \subseteq N_\varepsilon(\mathcal{A})$.

We now know that $\mathcal{A}$ is the limit of the Cauchy sequence $\{\mathcal{A}_n\}$, and we know from Lemma 9.2.4 that $\mathcal{A}$ is not empty. It remains to be shown that $\mathcal{A}$ is compact. By the Heine-Borel theorem (Theorem 6.2.1), this is equivalent to demonstrating that $\mathcal{A}$ is bounded and closed.

(c) $\mathcal{A}$ is bounded. We have seen in part (a) that, for given $\varepsilon > 0$ and sufficiently large $n, \mathcal{A} \subseteq N_\varepsilon(\mathcal{A}_n), \mathcal{A}_n \in K(S)$ and, hence, is compact. Consequently, $\mathcal{A}_n$ is bounded, i.e., there is $aR > 0$ such that $\mathcal{A}_n \subseteq N_R(0)$. If $y \in N_\varepsilon(\mathcal{A}_n)$, then $\rho(y, x) < \varepsilon$ for some $x \in \mathcal{A}_n$. Hence, $\rho(y, 0) \le \rho(0, x) + \rho(x, y) \le R + \varepsilon$. This is true for all $y \in N_\varepsilon(\mathcal{A}_n)$, and $N_\varepsilon(\mathcal{A}_n)$ stands revealed as bounded. Since $\mathcal{A}$ is contained therein, it is also bounded.

(d) $\mathcal{A}$ is closed, i.e., contains all its accumulation points (Lemma 6.1). Let $a \in S$ represent an accumulation point of $\mathcal{A}$. Then, there is a sequence $\{a_n\} \to a, a_n \in \mathcal{A}$. By the definition of $\mathcal{A}$, there is, for each a_n, a sequence $\{x_{nj}\} \to a_n$ with $x_{nj} \in \mathcal{A}_n$. Pick $N_1 < N_2 < N_3 < \cdots < N_j < \cdots$ so that $\rho(a_{N_j}, a) < 1/j$. For each N_j, there is an m_j such that $\rho(x_{N_j m_j}, a_{N_j}) < 1/j$. Hence, $\rho(x_{N_j m_j}, a) < 2/j$. Let $y_{N_j} = x_{N_j m_j}$. By construction, $y_{N_j} \in \mathcal{A}_{N_j}$

and $\lim_{j\to\infty} y_{N_j} = a$. We "fill up" this sequence, as in the proof of Lemma 9.2.4, to obtain a sequence $\{y_n\} \to a$ with $y_n \in \mathcal{A}_n$. Hence, $a \in \mathcal{A}$ and $\mathcal{A}$ is closed. □

9.3. The Invariant Attractor Set

To establish the existence and uniqueness of the attractor set, we will show that the mapping

$$\mathfrak{F}_*(\mathcal{A}) = \bigcup_{i=1}^{n} \mathfrak{F}_{i*}(\mathcal{A}), \tag{9.3.1}$$

where $\mathfrak{F}_i : S \to S, i = 1, 2, 3, \ldots, n$, are similarity transformations with reduction ratios $r_i \in (0, 1)$, is a contraction mapping. The contraction mapping theorem will take care of the rest.

(9.3.1) Theorem. *The mapping in (9.3.1) is a contraction mapping:* $d(\mathfrak{F}_*(\mathcal{A}), \mathfrak{F}_*(\mathcal{B})) \leq rd(\mathcal{A}, \mathcal{B})$, *where* $r = \max(r_1, r_2, \ldots, r_n) \in (0, 1)$.

Proof. Let $\mathcal{A}, \mathcal{B} \subset K(S)$ and choose some $\delta > d(\mathcal{A}, \mathcal{B})$. If $x \in \mathfrak{F}_*(\mathcal{A})$, then $x = \mathfrak{F}_i(x')$ for some $i \in \{1, 2, 3, \ldots, n\}$ where $x' \in \mathcal{A}$. Since $d(\mathcal{A}, \mathcal{B}) < \delta$, there is, by Lemma 9.2.3, a point $y' \in \mathcal{B}$ such that $\rho(x', y') < \delta$. $y = \mathfrak{F}_i(y') \in \mathfrak{F}_*(\mathcal{B})$. Hence, $\rho(x, y) \leq r\rho(x', y') < rd(\mathcal{A}, \mathcal{B}) < r\delta$. This is true for all $x \in \mathfrak{F}_*(\mathcal{A})$. Hence, $\mathfrak{F}_*(\mathcal{A}) \subseteq N_{r\delta}(\mathfrak{F}_*(\mathcal{B}))$. By symmetric reasoning, $\mathfrak{F}_*(\mathcal{B}) \subseteq N_{r\delta}(\mathcal{A})$. By Lemma 9.2.2, $d(\mathfrak{F}_*(\mathcal{A}), \mathfrak{F}_*(\mathcal{B})) \leq r\delta$. This is true for any $\delta > d(\mathcal{A}, \mathcal{B})$ and, hence, $d(\mathcal{F}_*(\mathcal{A}), \mathfrak{F}_*(\mathcal{B})) \leq rd(\mathcal{A}, \mathcal{B})$. □

The contraction mapping theorem in its various manifestations may be found in most respectable treatments of analysis. For the convenience of the reader, we will list it here, using our notation, and present a proof in the required setting.

(9.3.2) Theorem. *(Contraction mapping theorem). A contraction mapping* $\mathfrak{F}_* : K(S) \to K(S), d(\mathfrak{F}_*(\mathcal{A}), \mathfrak{F}_*(\mathcal{B})) \leq rd(\mathcal{A}, \mathcal{B}), 0 < r < 1$, *in the complete metric space* $K(S)$ *has a unique "fixed point"* $\mathcal{A} = \mathfrak{F}_*(\mathcal{A})$.

Proof. Let $\mathcal{A}_0 \in K(S)$ and $\mathcal{A}_{k+1} = \mathfrak{F}_*(\mathcal{A}_k), k = 0, 1, 2, 3, \ldots$. If $d(\mathcal{A}_0, \mathcal{A}_1) = \delta$, then

$$\begin{aligned} d(\mathcal{A}_1, \mathcal{A}_2) &= d(\mathfrak{F}_*(\mathcal{A}_0), \mathfrak{F}_*(\mathcal{A}_1)) \leq r\delta \\ d(\mathcal{A}_2, \mathcal{A}_3) &= d(\mathfrak{F}_*(\mathcal{A}_1), \mathfrak{F}_*(\mathcal{A}_2)) \leq r^2\delta \\ &\vdots \\ d(\mathcal{A}_n, \mathcal{A}_{n+1}) &= d(\mathfrak{F}_*(\mathcal{A}_{n-1}), \mathfrak{F}_*(\mathcal{A}_n)) \leq r^n\delta. \\ &\vdots \end{aligned}$$

Hence,

$$\begin{aligned} d(\mathcal{A}_m, \mathcal{A}_n) &\leq d(\mathcal{A}_m, \mathcal{A}_{m+1}) + d(\mathcal{A}_{m+1}, \mathcal{A}_{m+2}) + \cdots + d(\mathcal{A}_{n-1}, \mathcal{A}_n) \\ &\leq r^m\delta + r^{m+1}\delta + \cdots + r^{n-1}\delta = r^m\delta(1 + r + r^2 + \cdots + r^{n-1-m}) \\ &< \delta r^m/(1-r), \end{aligned}$$

and we see that $\{\mathcal{A}_n\}$ is a Cauchy sequence. Since $K(S)$ is complete, $\lim_{n\to\infty} \mathcal{A}_n = \mathcal{A} \in K(S)$ exists. We take the limit on both sides of $\mathcal{A}_{n+1} = \mathfrak{F}_*(\mathcal{A}_n)$ and obtain $\mathcal{A} = \mathcal{F}_*(\mathcal{A})$, meaning that a fixed point $\mathcal{A}$ exists. (Note that $\mathfrak{F}$ is continuous, because $d(\mathfrak{F}_*(\mathcal{A}), \mathfrak{F}_*(\mathcal{B})) \leq rd(\mathcal{A}.\mathcal{B}) < \varepsilon$ provided that $d(\mathcal{A}, \mathcal{B}) < \varepsilon/r$.)

To show that $\mathcal{A}$ is unique, we assume, to the contrary, that there are two distinct fixed points $\mathcal{A}$, $\mathcal{B}$. Then, $\mathcal{A} = \mathfrak{F}_*(\mathcal{A})$, $\mathcal{B} = \mathfrak{F}_*(\mathcal{B})$ and we have $d(\mathcal{A}, \mathcal{B}) = d(\mathfrak{F}_*(\mathcal{A}), \mathfrak{F}_*(\mathcal{B}) \leq rd(\mathcal{A}, \mathcal{B})$. Hence, $r \geq 1$, which contradicts our assumption that $0 < r < 1$. □

Theorems 9.3.1 and 9.3.2 yield:

(9.3.3) Theorem. *If S is a complete metric space and $\mathfrak{F}_i : S \to S$ represent similarity transformations with reduction ratios $r_i \in (0,1)$, the mapping $\mathfrak{F}_* : K(S) \to K(S)$, defined by $\mathfrak{F}_*(\mathcal{S}) = \bigcup_{i=1}^{n} \mathfrak{F}_{i*}(\mathcal{S}), \mathcal{S} \in K(S)$, has a compact attractor set $\mathcal{A} = \mathfrak{F}_*(\mathcal{A})$ and this attractor set is unique.*

We see from Theorem 9.3.3 that the attractor set is independent of the choice for the initial set $\mathcal{A}_0$. For example, had we started out with an initial set other than $\mathcal{I}$ when generating the Cantor set in Section 9.1 by means of the iterated function system (9.1.1), we would still have obtained the Cantor set, as long as the initial set was a non-empty compact subset of $\mathbb{E}^1$. Suppose we had started out with the interval [1,3/2]. Application of (9.1.1) puts the beginning point after one iteration at 1/3, and then $1/9, 1/27, 1/81, \ldots \to 0$. The endpoint 3/2 would have moved to $3/6+2/3 = 7/6$, and then to $19/18, 55/54, \ldots \to 1$, etc....

$\mathcal{Q}$ is the attractor set of the mappings defined in (2.4.1) (Hilbert curve) and (3.4.1) (Peano curve), $\mathcal{T}$ is the attractor set of the mapping in (4.3.1) (Sierpiński-Knopp curve), and W is the attractor set of the mapping in (2.8.2) (three-dimensional Hilbert curve). We now know that we would have obtained these attractor sets, no matter which compact set we had chosen as the initial set. However, to obtain meaningful intermediate results, namely, approximating curves, one would want to choose as an initial set (Leitmotiv) a curve that connects the entry point to the exit point.

One can show that, if the attractor set $\mathcal{A}$ of an iterated function system $x' = \mathfrak{F}_k(x)$, such as we have considered in this and the preceding two sections, is pathwise connected (it is compact by virtue of Theorem 9.3.3), and there are distinct points $(\xi_k, \eta_k) \in \mathcal{A}, k = 0, 1, 2, \ldots, n$, such that

$\mathfrak{F}_k(\xi_0, \eta_0) = (\xi_{k-1}, \eta_{k-1}), \mathfrak{F}_k(\xi_n, \eta_n) = (\xi_k, \eta_k), k = 1, 2, 3, \ldots, n$ (which assures us that the exit point from $\mathfrak{F}_{k-1*}(\mathcal{A})$ coincides with the entry point into $\mathfrak{F}_{k*}(\mathcal{A})$), then there is a continuous function $f : [0,1] \overset{\text{onto}}{\longrightarrow} \mathcal{A}$ which interpolates the points $(\xi_0, \eta_0), (\xi_1, \eta_1), \ldots, (\xi_n, \eta_n)$, with (ξ_0, η_0) representing the beginning point and (ξ_n, η_n) the endpoint: $f(0) = (\xi_0, \eta_0), f(1) = (\xi_n, \eta_n)$. If $J_2(\mathcal{A}) > 0$, then f represents a space-filling curve (Barnsley [1], pp. 240–245).

9.4. Similarity Dimension

When comparing the von Koch curve with the Sierpiński gasket, one feels intuitively that the latter is of greater complexity and occupies the plane more densely. Similarly, the modified Sierpiński triangle appears more complex than the Sierpiński gasket. However, neither their Lebesgue measure nor their topological dimension offers a clue to this. We have already seen in Section 8.3 that the von Koch curve has Lebesgue measure zero. When constructing the Sierpiński gasket, starting out with an equilateral triangle of area A, we remove, at the first step, a triangle of area A/4. At the next step, three triangles of area A/16 each, and, at the next step, nine triangles of area A/64 each, etc..., to be left with a set of Lebesgue measure $0 : \Lambda_2(\textit{Sierpiński triangle}) = 0$. This should not come as a surprise because, whenever in the construction of a fractal, at every step, the content of the trema stands at a fixed ratio $\rho \in (0,1)$ to the content of the portion from which it is removed, the Lebesgue measure of what is ultimately left is zero: Suppose that the content of the initial set is 1. After one step, $(1-\rho)$ is left. After two steps, $1 - \rho - \rho(1-\rho) = (1-\rho)^2$ is left,... after n steps, $(1-\rho)^n$ is left, and $(1-\rho)^n \to 0$ as $n \to \infty$. (See also Section 8.3.)

When constructing the modified Sierpiński triangle, we dissected the initial triangle into nine congruent triangles (see Fig. 9.1.4), eliminated three triangles, and retained six. The area of the trema stands in the fixed ratio 1/3 to the area from which it is removed. Hence, $\Lambda_2(\textit{modified Sierpiński triangle}) = 0$.

The topological dimension is equally unrevealing. Since we will only use it as an illustration, we give here a very informal definition: We say that a set has topological dimension zero if no point in the set can move. For example, a set that contains finitely many points has topological dimension zero because none of its points can move without leaving the set. A set has topological dimension one if any point in the set can be trapped by a subset of dimension zero. For example, a line segment has dimension one because any point in it can be trapped by a set containing two points only which has, by the preceding definition, topological dimension zero. A Jordan curve is another example of a set of topological dimension one. A set has topological dimension two if any point in the set can be trapped by a subset of topological dimension one. For example, a square has topological dimension two because any point can be trapped by a closed Jordan curve (which has, by the preceding Definition, topological dimension one). This suffices to explain the inductive process in the definition of topological dimension. It is now clear that the von Koch curve, the Sierpiński gasket, and the modified Sierpiński triangle have topological dimension one and that the topological dimension will not enable us to differentiate between the complexity and density of these curves.

It is, however, possible (in more ways than one) to assign non-integral dimensions to sets that will provide a more accurate measure of their complexity and the density with which they occupy the space than either the topological dimension or the Lebesgue measure. Here, we will only deal with self-similar fractals, which allow for the simplest concept of non-integral dimensions. We motivate the upcoming definition as follows: A one-dimensional line segment can be broken up into n equal parts of length $1/n$ each. A two-dimensional set, when scaled down and copied n times, has to be scaled down by the factor $1/\sqrt{n}$, if the n congruent non-overlapping copies are to occupy the same amount of space as the original set. If a three-dimensional set is copied n times, then it will have to be scaled down by the factor $1/\sqrt[3]{n}$ for the n congruent non-overlapping copies to occupy the same amount of space as the original set. In general, if an s-dimensional set is copied n times and the n non-overlapping copies are to occupy the same amount of space as before, it has to be scaled down by the factor $r = 1/\sqrt[s]{n}$. This gives rise to the following definition:

(9.4) Definition. *A self-similar fractal (attractor set) that is generated by a catalogue of n similarity transformations $x' = \mathfrak{F}_j(x)$, $j = 1, 2, 3, \ldots, n$, with reduction ratio $r \in (0, 1)$ is said to have a similarity dimension of*

$$s = \frac{\log(n)}{\log(1/r)}$$

(where it obviously does not matter which logarithm is used). More generally, if $\mathfrak{F}_1$ has reduction ratio $r_1 \in (0, 1)$, $\mathfrak{F}_2$ reduction ratio $r_2 \in (0, 1), \ldots$, and $\mathfrak{F}_n$ reduction ratio $r_n \in (0, 1)$, then the similarity dimension s of the attractor set is the (unique) solution of

$$r_1^s + r_2^s + \cdots + r_n^s = 1.$$

Note that this definition is not tied to the attractor set itself but to the iterated function system that generates it. It is conceivable that the same attractor set is generated by different iterated function systems, leading to different similarity dimensions. For example, the iterated function system

$$\xi' = \frac{\xi}{2}, \xi' = \frac{\xi + 1}{2}$$

has the interval $\mathcal{I}$ as attractor set and, by Definition 9.4, the similarity dimension of $\mathcal{I}$ appears to be $\log 2/\log 2 = 1$. But, $\mathcal{I}$ is also the attractor set of

$$\xi' = \frac{2\xi}{3}, \xi' = \frac{2\xi + 1}{3}$$

and now the similarity dimension of $\mathcal{I}$ appears to be $\log 2/\log(3/2) \cong 1.709511$. In the latter case, there is overlap of the interior of the images and these overlapping portions are "counted" twice. One can show that this cannot happen if there is an open set $\mathcal{U}$ such that $\mathfrak{F}_{i*}(\mathcal{U}) \subseteq \mathcal{U}$ for all $i = 1, 2, 3, \ldots, n$, and $\mathfrak{F}_{i*}(\mathcal{U}) \cap \mathfrak{F}_{j*}(U) = \emptyset$ for $i \neq j$, which prevents interior parts from overlapping. (This is called *Moran's open set condition.)* In such a case, the similarity dimension is equal to the *Hausdorff dimension*, a concept that is tied to the set itself and not to the process that generates it (Barnsley [1], Chapter 5, Edgar [1], Chapter 6). The Hausdorff dimension, incidentally, is always greater than or equal to the topological dimension and less than or equal to the similarity dimension (Edgar [1], p. 156).

We are now in a position to compute the similarity dimension of the fractals of Section 9.1.

Cantor set: In generating the Cantor set, at each step, the preceding result is scaled down in the ratio $r = 1/3$, and there are $n = 2$ similarity transformations, making two scaled-down copies of the preceding configuration at every step of the iteration. Hence,

$$s(\textit{Cantor set}) = \frac{\log(2)}{\log(3)} \cong 0.6309296.$$

(If we take $\mathcal{U} = (0, 1)$, then $\mathfrak{F}_{1*}(\mathcal{U}) = (0, 1/3), \mathfrak{F}_{2*}(\mathcal{U}) = (2/3, 1)$. These two sets are contained in $\mathcal{U}$ and are disjoint. Hence, the Hausdorff dimension of the Cantor set is also $\log(2)/\log(3)$.) The topological dimension of the Cantor set is, by contrast, 0, because the Cantor set does not contain any open intervals and, hence, none of its points can move.

von Koch curve: At each step, the preceding configuration is scaled down in the ratio 1/3 and $n = 4$ copies are made:

$$s(\textit{von Koch curve}) = \frac{\log(4)}{\log(3)} \cong 1.261859.$$

(If we take $\mathcal{U}$ to be the interior of the initial triangle in Fig. 8.3.3, we see again that the similarity dimension and the Hausdorff dimension are the same.) The von Koch curve is a Jordan curve with topological dimension one.

Sierpiński gasket: At each step, the configuration is scaled down in the ratio 1/2 and $n = 3$ copies are made:

$$s(\textit{Sierpiński gasket}) = \frac{\log(3)}{\log(2)} \cong 1.584963.$$

(To check whether the similarity dimension and the Hausdorff dimension are the same, take $\mathcal{U}$ to be the interior of the initial triangle.) The topological dimension of the Sierpiński gasket is one.

Modified Sierpiński triangle: The reduction ratio is 1/3 and we retain, at every step, six copies. Hence, the similarity dimension is given by

$$s(\textit{Modified Sierpiński triangle}) = \frac{\log(6)}{\log(3)} \cong 1.63093.$$

Again, the reader can be easily convinced that the similarity dimension is equal to the Hausdorff dimension.

We see now that we were not misled by our intuition. The similarity dimension of the Sierpiński gasket is indeed greater than that of the von Koch curve, and the similarity dimension of the modified Sierpiński triangle is, in turn, greater than the similarity dimension of the Sierpiński gasket.

Hilbert curve and *Peano curve:* In the first case, we scale down in the ratio 1/2 and make four copies, and in the case of the Peano curve we scale down in the ratio 1/3 and make nine copies. Hence, to no surprise,

$$s(\textit{Hilbert curve, Peano curve}) = \frac{\log(1/r^2)}{\log(1/r)} = 2.$$

(Again, the similarity dimension and the Hausdorff dimension are the same, as one can see by taking the interior of the initial square as $\mathcal{U}$.) In this case, the topological dimension is also two. So, all three dimensions coincide.)

Sierpiński curve: We scale down in the ratio $1/\sqrt{2}$ and make two copies:

$$s(\textit{Sierpiński curve}) = \frac{\log 2}{\log \sqrt{2}} = 2.$$

(To check the Moran condition, take for $\mathcal{U}$ the interior of the initial triangle. As in the preceding case, the similarity dimension, the Hausdorff dimension, and the topological dimension are the same.)

Purists only use the term *fractal* for sets with non-integral dimension. So, the Hilbert-type space-filling curves would not be considered fractals. At the other extreme, there are those who consider every non-empty compact subset of a complete metric space a fractal. Efforts are being made to define a fractal. A pursuit of this question would lead us far afield and properly belongs to a treatise on fractals, where one may occasionally find it.

9.5. Cantor Curves

A most interesting self-similar fractal is the *square carpet*, also referred to as the *Sierpiński carpet* or the *Sierpiński continuum* (Sierpiński [6]). Sierpiński [5] states in his last paragraph that Mazurkiewicz had already found this example but, for reasons unknown to him, never published it (*"Jusqu'à présent M. Mazurkiewicz ne publia pas son exemple et sa démonstration m'est inconnue"*—see also Sierpiński [7], p. 632). We will call it the *Mazurkiewicz continuum.* We start out with the unit square, dissect it into nine congruent squares, and remove the middle square. Then we remove the middle squares of the remaining eight squares, etc.... The first two steps are illustrated in Fig. 9.5.1. The fifth iteration, depicted in Fig. 9.5.1a, was produced by N.J. Rose on an Apple Macintosh computer, using his own program. The reduction ratio is 1/3 and eight copies are made at every step. Hence,

$$s(\textit{Mazurkiewicz continuum}) = \frac{\log(8)}{\log(3)} \cong 1.892789.$$

(Since Moran's open set condition is obviously met with the interior of the square as $\mathcal{U}$, the Hausdorff dimension is the same.)

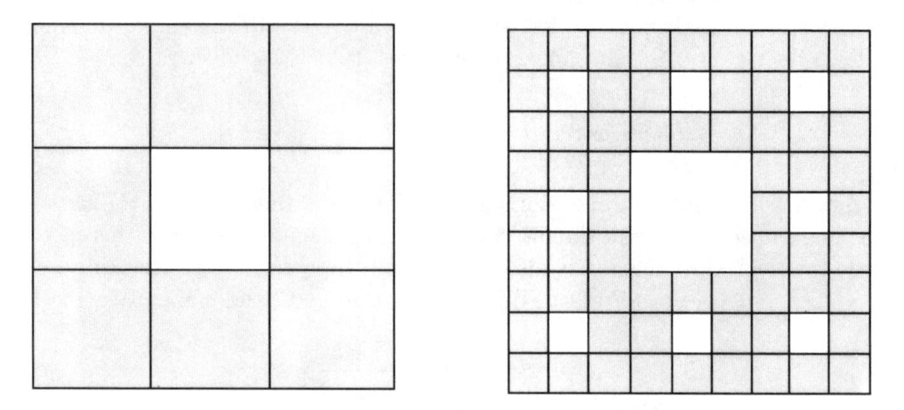

Fig. 9.5.1. Generating the Mazurkiewicz Continuum

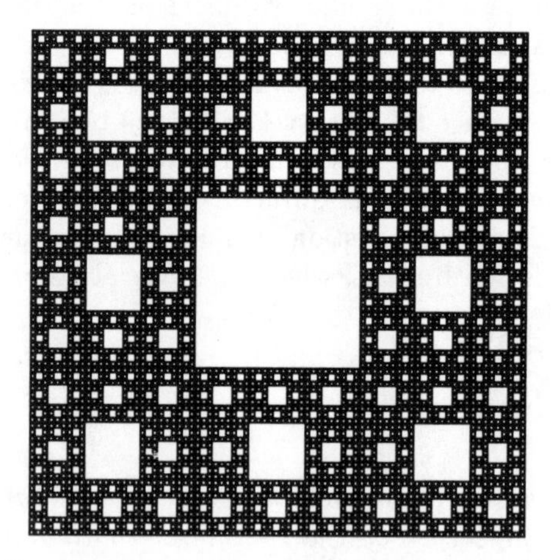

Fig. 9.5.1a. Fifth Iteration in the Generation of the Mazurkiewicz Continuum

The two-dimensional Lebesgue measure of the *Mazurkiewicz continuum* is zero because, again, the ratio of the content of the trema to the content of the set from which it is removed is constant.

To explain the significance of the Mazurkiewicz continuum within the context of the theory of plane curves, let us first state what is meant by a *Cantor curve*: A Cantor curve is a *continuum* (a compact connected set that contains at least two points) in $\mathbb{E}^2$, that is nowhere dense in $\mathbb{E}^2$. (A set is nowhere dense in $\mathbb{E}^2$ if none of its subsets is dense in any open subset of $\mathbb{E}^2$.) Clearly, a Cantor curve cannot contain a square. Hence, space-filling curves are not Cantor curves. On the other hand, the set in Fig. 6.3.1 is a Cantor curve. This is in sharp contrast to our Definition 1.3.2, by which a square is a curve but the set in Fig. 6.3.1 is not. It turns out that each Cantor curve is homeomorphic to a subset of the Mazurkiewicz continuum. Because of this, the Mazurkiewicz continuum is often called the *universal plane curve* (Blumenthal and Menger [1], p. 431).

Again, we leave it to the reader to find the appropriate iterated function system that generates the Mazurkiewicz continuum.

Let us now consider a three-dimensional generalization of the Mazurkiewicz continuum (which B.B. Mandelbrot [1] refers to as the *Sierpiński sponge*), which is also known as the *Menger sponge*. It is obtained from the unit cube $\mathcal{W}$ by partitioning it into 27 congruent subcubes and, at the first step, removing the center cube and the center cubes of all six faces. At the next step, we subject each of the remaining 20 cubes to the same treatment, etc.... (See Fig. 9.5.2, which is taken from Blumenthal and Menger [1] by permission from the publishers, W.H. Freeman and Company).

Since the reduction ratio is 1/3, and since 20 copies are made at every step, the similarity dimension is given by

$$s(\textit{Sierpiński sponge}) = \frac{\log(20)}{\log(3)} = 2.726833.$$

Since Moran's open set condition is obviously met with $\mathcal{U}$ as the interior of $\mathcal{W}$, the Hausdorff dimension is the same as the similarity dimension. However, the topological dimension of the Sierpiński sponge is only one. In fact, each Cantor curve in $\mathbb{E}^3$ is homeomorphic to a subset of the Sierpiński sponge. A Cantor curve $\mathcal{C}$ in $\mathbb{E}^3$ is a continuum with the property that for each point p, each neighborhood of p contains a neighborhood $\mathcal{N}$ with the property that the intersection of $\mathcal{C}$ with the set of all accumulation points of $\mathcal{N}$ that do not lie in $\mathcal{N}$ does not contain a sub-continuum of $\mathcal{C}$ (Blumenthal and Menger [1], p. 439). This is a generalization of the concept of Cantor curves in $\mathbb{E}^2$ which we defined earlier. In $\mathbb{E}^2$, the two concepts coincide (Blumenthal and Menger [1], pp. 481, 493. Note that a direct generalization of the earlier definition of Cantor curves to $\mathbb{E}^3$ would have led to a concept that embraces two-dimensional surfaces.) Since every Cantor curve in a metric space is homeomorphic to a Cantor curve in $\mathbb{E}^3$, it follows that

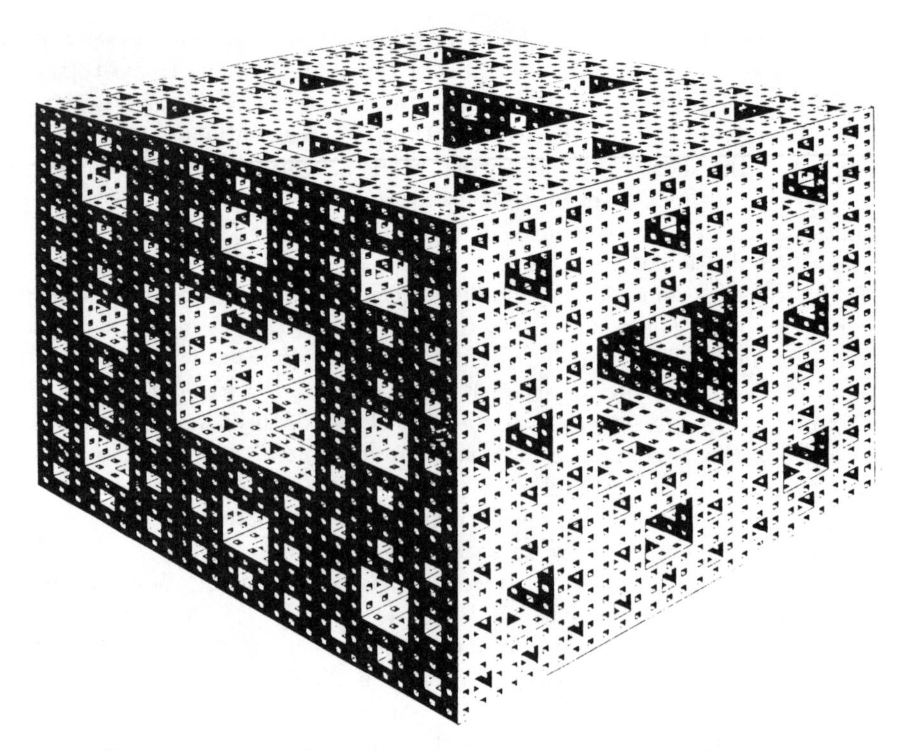

Fig. 9.5.2. Fourth Step in the Generation of the Sierpiński Sponge

every Cantor curve is homeomorphic to a subset of the Sierpiński sponge (Blumenthal and Menger [1], pp. 501–506). Because of this, the Sierpiński sponge is also referred to as *the universal curve*. Since a Jordan curve is a Cantor curve, it follows, in particular, that every Jordan curve is homeomorphic to a subset of the Sierpiński sponge. As in the previous cases where the ratio of the content of the trema to the content of the set, it is removed from was constant we have, for the three-dimensional Lebesgue measure:

$$\Lambda_3(\textit{Sierpiński sponge}) = 0.$$

We leave it to the reader to set up the appropriate iterated function system.

9.6. The Heighway Dragon

Around 1967, John E. Heighway, a physicist at the University of Irvine, California, proposed at a NASA seminar on group theory the construction of a most remarkable fractal, which he named a dragon (see Gardner [1], Edgar [1], and Davis and Knuth [1], [2]). Hence the name Heighway dragon.

The Heighway dragon may be obtained by applying the following iterated function system

$$
\begin{aligned}
\begin{pmatrix}\xi'\\ \eta'\end{pmatrix} &= \frac{1}{2}\begin{pmatrix}0 & 1\\ -1 & 0\end{pmatrix}\begin{pmatrix}\xi\\ \eta\end{pmatrix}\\
\begin{pmatrix}\xi'\\ \eta'\end{pmatrix} &= \frac{1}{2}\begin{pmatrix}-1 & 0\\ 0 & -1\end{pmatrix}\begin{pmatrix}\xi\\ \eta\end{pmatrix}+\frac{1}{2}\begin{pmatrix}1\\ -1\end{pmatrix}\\
\begin{pmatrix}\xi'\\ \eta'\end{pmatrix} &= \frac{1}{2}\begin{pmatrix}0 & -1\\ 1 & 0\end{pmatrix}\begin{pmatrix}\xi\\ \eta\end{pmatrix}+\frac{1}{2}\begin{pmatrix}1\\ -1\end{pmatrix}\\
\begin{pmatrix}\xi'\\ \eta'\end{pmatrix} &= \frac{1}{2}\begin{pmatrix}-1 & 0\\ 0 & -1\end{pmatrix}\begin{pmatrix}\xi\\ \eta\end{pmatrix}+\frac{1}{2}\begin{pmatrix}2\\ 0\end{pmatrix}
\end{aligned}
\tag{9.6.1}
$$

to the interval $\mathcal{I}$. The first five steps are illustrated in Fig. 9.6.1. We call the initial polygon, namely $\mathcal{I}, \mathcal{P}_0$, the next one $\mathcal{P}_1$, etc. . . .
Note how all approximating polygons emanate from (0,0) and terminate at (1,0). (We owe the drawings in Fig. 9.6.2 to Nicholas J. Rose, who generated them on an Apple Macintosh computer using his own program.)

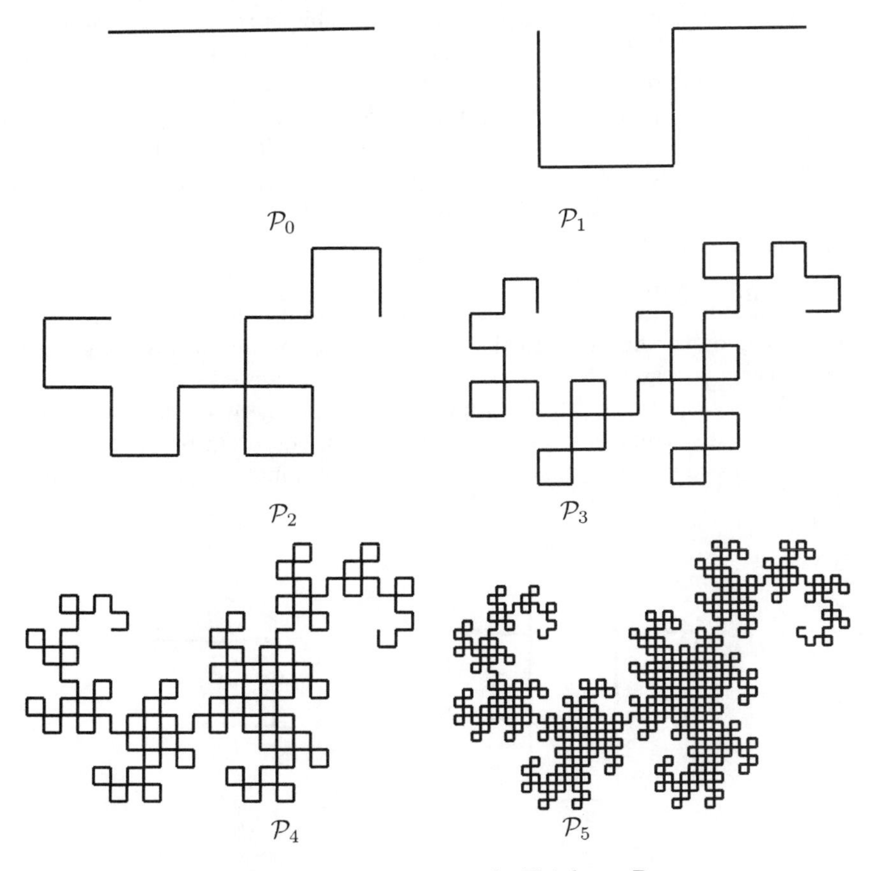

Fig. 9.6.1. Approximations to the Heighway Dragon

With the reduction ratio of 1/2 and with four copies being made at every step, we obtain the similarity dimension

$$s(\text{Heighway dragon}) = \frac{\log(4)}{\log(2)} = 2.$$

The fact that the similarity dimension of the Heighway dragon is two does not in any way guarantee that we are dealing with a space-filling curve. However, it does make it a prime candidate, and, as the following argument will reveal, it is indeed space-filling, inasmuch as it contains at least one closed square of sidelength 1/16.

$\mathcal{P}_4$ contains at least one configuration, as depicted in Fig. 9.6.2(a) (with the center square represented by $1/2 \leq \xi \leq 9/6$, $-1/4 \leq \eta \leq -3/16$). Let us call the square in the center $\mathcal{Q}_4$. In $\mathcal{P}_5$, $\mathcal{Q}_4$ appears again, partitioned into four congruent subsquares, each of which is the central square of a configuration such as in Fig. 9.6.2(a). This is depicted in Fig. 9.6.2(b). Now, each of these four central squares will appear again in $\mathcal{P}_6$, where they are dissected into four congruent subsquares, each one having a square of the same size attached to each of its sides. This is repeated ad infinitum. Let x represent a point in Q_4 of sidelength 1/16, and let $x_4 \in \partial Q_4$ ($\partial \mathcal{Q}_j$ represents the boundary of the set $\mathcal{Q}_j$). x lies in one (or more) of the four subsquares of Q_4. Call it $\mathcal{Q}_5$. $\mathcal{Q}_5$ has sidelength 1/32. Let $x_5 \in \partial \mathcal{Q}_5$. x lies in one (or more) of the four subsquares of $\mathcal{Q}_5$; call it Q_6. Q_6 has sidelength 1/64. Let $x_6 \in \partial \mathcal{Q}_6$. We continue this process ad infinitum. By construction, we have $x_j \in \partial \mathcal{Q}_j \subset \mathcal{A}$, where $\mathcal{A}$ denotes the attractor set, and $\lim_{n\to\infty} x_j = x$. By Theorem 9.3.3, $\mathcal{A}$ is closed and, hence, $x \in \mathcal{A}$. Since x was any point in Q_4, we see that $Q_4 \subset \mathcal{A}$. □

The Heighway dragon seems to be the latest original space-filling curve to appear on the mathematical scene, not counting modifications, generalizations, or specializations of known space-filling curves.

As was the case with the approximating polygons to the space-filling curves we discussed in Chapters 2 through 5, the progress of the dragon curve is obscured by the fact that it keeps bumping into itself. To exhibit

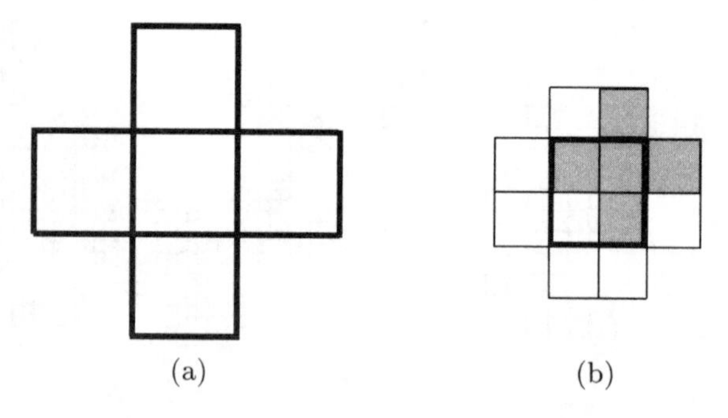

Fig. 9.6.2. Parts of $\mathcal{P}_4$ and $\mathcal{P}_5$

Fig. 9.6.3. $\mathcal{P}_6$ with Rounded Corners

that it never crosses itself, and never doubles up, one may round off the corners. Figure 9.6.3 (which is a copy of Fig. 4 in Davis and Knuth [1], p. 69 and is reproduced here by permission from the authors) depicts $\mathcal{P}_6$ with rounded corners. Figure 9.6.3 also reveals where the name dragon curve came from. If the von Koch curve is, as Voss said, one of the early mathematical monsters, then the Heighway dragon is one of the latest!

9.7. Problems

1. Show that the Cantor set does not contain any open intervals.
2. Let U denote an orthogonal $m \times m$ matrix and let $x' = rUx + b$, where $x, x', b \in \mathbb{E}^m$, $r \in \mathbb{R}$. Show that $||x'_1 - x'_2|| = r||x_1 - x_2||$.
3. Apply the iterated function system (9.1.1) to the initial set $\mathcal{A}_0 = \{\xi \in \mathbb{R} | \xi = 2\}$ and take it through several iterations. Sketch the intermediate results.
4. Repeat Problem 3 for the iterated function system (9.1.2) and the initial set $\mathcal{A}_0 = \mathcal{T}$.
5. Repeat Problem 3 for the iterated function system (9.3.1) and the initial set $\mathcal{A} = \{(\xi, \eta) \in \mathbb{E}^2 | \eta = \sin(\pi\xi/2), 0 \leq \xi \leq 2\}$.
6. Prove Lemma 9.2.2.
7. Show: If $r_1^s + r_2^s + \cdots + r_n^s = 1$ and $r_1 = r_2 = \cdots = r_n = r$, then $s = \log(n)/\log(1/r)$.
8. Show that $r_1^s + r_2^s + \cdots + r_n^s = 1, r_i \in (0, 1)$ has a unique solution s.

9. Find the similarity dimension of the Pólya curve of Section 4.6.
10. Let $\mathcal{A}_n = \{x \in S | \rho(1/n, x) \leq 1\} \in K(S)$. Find $\lim_{n\to\infty} \mathcal{A}_n$.
11. Let $\mathcal{A}_n = \{x \in S | \rho(0, x) \leq 1 + (1/n)\} \in K(S)\}$. Find $\lim_{n\to\infty} \mathcal{A}_n$.
12. Find the iterated function system that generates the modified Sierpiński triangle of Section 9.1.
13. Repeat Problem 12 for the Mazurkiewicz continuum (square carpet) of Section 9.5.
14. Repeat Problem 12 for the Sierpiński sponge of Section 9.5.
15. Find the iterated function system that maps $\mathcal{I}$ onto the configuration in Fig. 9.7. Draw several iterates and compute the similarity dimension of this fractal (also called the quadratic von Koch curve).

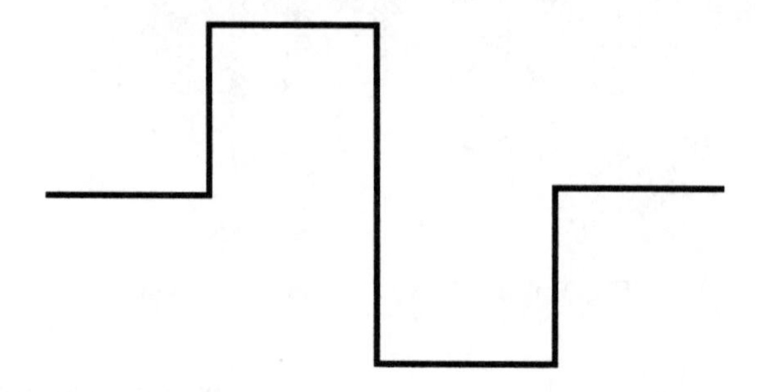

Fig. 9.7. Second Iterate of the Quadratic von Koch curve

16. The iterated function system (9.6.1) cannot be used to parametrize the Heighway dragon by the method employed in Section 2.3 to parametrize the Hilbert curve, and in Section 4.3 to parametrize the Sierpiński curve. Explain why.
17. Is the Cantor brush of Problem 6.26 a Cantor curve? Justify your answer.
18. Show that the iterated function system

$$x' = \begin{pmatrix} 1/3 & 0 \\ 0 & 1/2 \end{pmatrix} x,$$

$$x' = \begin{pmatrix} 1/3 & 0 \\ 0 & 0 \end{pmatrix} x + \begin{pmatrix} 1/3 \\ 1/2 \end{pmatrix},$$

$$x' = \begin{pmatrix} 1/3 & 0 \\ 0 & 1/2 \end{pmatrix} x + \begin{pmatrix} 2/3 \\ 1/2 \end{pmatrix},$$

when applied to $\mathcal{I}$, generates the Devil's Staircase of Section 5.3. (This is a special case of a Kieswetter curve. For more details, see Kuttler [1].)
19. The iterated function system

$$x' = \frac{1}{2}\begin{pmatrix} 1 & 0 \\ 1 & 1 \end{pmatrix} x, \qquad x' = \frac{1}{2}\begin{pmatrix} 1 & 0 \\ -1 & 1 \end{pmatrix} x + \frac{1}{2}\begin{pmatrix} 1 \\ 1 \end{pmatrix}$$

generates the Takagi curve. Apply to $\mathcal{I}$ and sketch a few iterations.

20. Repeat Problem 19 for the iterated function system

$$x' = \frac{3}{8}\begin{pmatrix} 1 & 0 \\ 0 & 2 \end{pmatrix} x$$
$$x' = \frac{1}{8}\begin{pmatrix} -1 & 0 \\ 0 & 2 \end{pmatrix} x + \frac{1}{2}\begin{pmatrix} 1 \\ 1 \end{pmatrix}$$
$$x' = \frac{3}{8}\begin{pmatrix} 1 & 0 \\ 0 & 2 \end{pmatrix} x + \frac{1}{2}\begin{pmatrix} 1 \\ 1 \end{pmatrix}$$
$$x' = \frac{1}{8}\begin{pmatrix} -1 & 0 \\ 0 & 2 \end{pmatrix} x + \frac{1}{2}\begin{pmatrix} 2 \\ 2 \end{pmatrix},$$

which generates the Bolzano-Kowalewski curve.

Appendix

A.1. Computer Programs

In the following three programs to be used for the computation of the coordinates of the nodal points of the Hilbert curve, the Peano curve, and the Sierpiński-Knopp curve, we enter the value of the independent variable $t \in [0,1]$ as a string variable. This is because the computations for the Hilbert Curve and the Sierpiński-Knopp curve require an input in quaternary form and for the Peano curve in ternary form, but the computer does not recognize inputs in such a form unless we include a conversion subroutine, which would complicate the program and make it more sluggish. We will then peel off one digit at a time, and convert it to its numerical value. In the program for the Peano curve, the output is also in ternary form and we will therefore use string variables also for the output. To keep the programs simple, we have not included any "fail-safe" devices. While an illegitimate input will not be rejected, it will only produce a nonsensical result.

A.1.1. Computation of the Nodal Points of the Hilbert Curve

Program 1 in IBM PC Personal Computer BASIC produces the coordinates of the nodal points of the Hilbert curve by evaluating the formula (2.4.3). The variable t is entered in the form $0.t_1t_2t_3\ldots t_n$, where the point after the leading 0 is to be understood as a quaternary point and where $n \leq 253$. With double precision, the output which is in the form of a fraction with the value of 2^n in the denominator, will be precise as long as $n \leq 53$.

Program 1
(for the evaluation of Formula (2.4.3))

```
10   DEFDBL F,G,W
20   K=0:D=0:Q=0:E0=0:E3=0:F=0:G=0
30   PRINT ''Enter t in quaternary form (such as,
       for example, 0.0312)''
40   INPUT X$
50   N=LEN (X$)-2
```

```
 60   Q$ = MID$ (X$,K+3,1)
 70   Q=VAL (Q$)
 80   K=K+1:N=N-1
 90   F=F+(-1)^E0 * ((1-D)*Q-1)*SGN(Q)*2^N
100   G=G+(-1)^E0 * (1-D*Q)*SGN(Q)*2^N
110   IF N=0 THEN 200
120   IF Q=0 THEN 150
130   IF Q=3 THEN 170
140   GOTO 180
150   E0=(E0+1) MOD 2
160   GOTO 180
170   E3=(E3+1) MOD 2
180   D=E0+E3 MOD 2
190   GOTO 60
200   W=2^K
210   PRINT ``f(``+X$ +'') = ``;F;''/'';W
220   PRINT ``g(``+X$ +'') = ``;G;''/'';W
```

In the above program, Q stands for q_j, E0, E3 for e_{0j}, e_{3j}, D for d_j, and F, G for φ_h, ψ_h.

A plotting program for the approximating polygons of Figs. 2.1.1 and 2.1.2 may be found in Wirth ([1], pp. 132, 133).

A.1.2. Computation of the Nodal Points of the Peano Curve

Program 2 in IBM PC Personal Computer BASIC produces the coordinates of the nodal points of the Peano curve by evaluating (3.1.1). The variable t is entered in the form $t = 0.t_1t_2t_3 \ldots t_n$, where the point after the leading 0 represents a ternary point, and where $n \leq 253$. Since input and output are both handled as string variables, the output will be precise as long as $n \leq 253$. Observe that we used the same routine (lines 60–140) for the computation of the components φ_p and ψ_p. We could do this because both are calculated by the same procedure. In the computation of the first component φ_p we start operating on the first digit after the ternary point (K = 0) and in the computation of the second component ψ_p, with the second digit after the ternary point (K=1 in line 170). Also, note that VAL(B$) = 0 for B$ = "." in 60 for K=0.

Program 2
(for the evaluation of Formula (3.1.1))

```
 10   PRINT ``Enter t in ternary form (such as,
        for example, 0.2102)''
 20   INPUT X$:Z$=X$
 30   B=0:F$=``0 .'':I=I+1
 40   N=LEN (Z$): M=N MOD 2:IF M=I MOD 2 THEN 60
 50   N=N+1:X$=Z$ + STR$(0)
```

```
 60   B$=MID$(X$,K+2,1):B=B+VAL(B$)
 70   A$=MID$(X$,K+3,1):A=VAL(A$)
 80   E=B MOD 2:IF E=0 THEN 100
 90   A=2-A
100   F$=F$ +STR$(A)
110   K=K+2:N=N-2:IF N<=I THEN 130
120   GOTO 60
130   IF E=0 THEN 150
140   F$=F$ + '' 2 2 2. . .''
150   ON I GOTO 160, 180
160   PRINT ''f(''+Z$ +'') = ''+F$:X$=Z$
170   K=1:GOTO 30
180   PRINT ''g(''+Z$ +'') = ''+F$
```

In the above program, A stands for $t_{2j+1}, j = 0, 1, 2, \ldots$ and B stands for t_{2j}, $j = 0, 1, 2, \ldots$ with t_0 = VAL(".") = 0. F, G stand for φ_p, ψ_p.

A.1.3. Computation of the Nodal Points of the Sierpiński-Knopp Curve

Program 3 in IBM PC Personal Computer BASIC produces the coordinates of the nodal points of the Sierpiński-Knopp curve by evaluating formula (4.3.4). The variable t is entered in the form $t = 0.q_1q_2q_3 \ldots q_n$, where $n \leq 253$, and where the point after the leading 0 is to be taken as a quaternary point. The result is precise as long as $n \leq 54$, which is slightly better than in Program 1.

Program 3
(for the evaluation of Formula (4.3.4))

```
 10   DEFDBL F,G,W
 20   K=0:D=0:Q=0:E=0:F=0:G=0
 30   PRINT ''Enter t in quaternary form (such as,
        for example, 0.0312)''
 40   INPUT X$
 50   N=LEN(X$)-2
 60   Q$=MID$(X$,K+3,1)
 70   Q=VAL(Q$)
 80   K=K+1:N=N-1
 90   F=F+(-1) ^ E*((1-D)*(1+(-1) ^ D)+(D-2)*(1+(-1)
        ^ Q)*(1-(-1) ^ D)/2)* SGN (Q)*2 ^ (N-1)
100   G=G+(-1) ^ E*((2-D)*(1-(-1) ^ D)+(1-D)*(1+(-1)
        ^ Q)*(1+(-1) ^ D)/2)* SGN (Q)*2 ^ (N-1)
110   IF N=0 THEN 180
120   IF Q=2 THEN 150
130   IF Q=1 THEN 160
140   GOTO 170
150   E=(E+1) MOD 2
```

```
160    D=(D+1) MOD 4
170    GOTO 60
180    W=2 ^ (K-1)
190    PRINT ''f(''+X$ +'') = '';F;''/'';W
200    PRINT ''g(''+X$ +'') = '';G;''/'';W
```

In the above program, Q stands for q_j, E for η_j, D for δ_j, and F, G for φ_s, ψ_s.

A plotting program for the approximating polygons of the original Sierpiński curve of Fig. 4.1 may be found in Wirth ([1], pp. 136, 137).

A.1.4. Plotting Program for the Approximating Polygons of the Schoenberg Curve

The following two programs produce the graphs of the s_n-and the a_n-approximating polygons to the Schoenberg curve. The s_n-approximating polygons are defined in (7.3.2) and the a_n-approximating polygons in (7.3.4).

Program 4
(generating the graphs of the s_n approximations)

```
 10    DEFDBL T,X,Y
 20    CLS:PRINT:Enter n = '';:INPUT N:SCREEN 2:
         PSET (0,198):M=1
 30    DEF FNA(T)=(3*T-1)*(SGN(SGN(3*T-1)+1)
         -SGN(SGN(3*T-2)+1))
 40    DEF FNB(T)=SGN(SGN(3*T-2)+1)-SGN(SGN(3*T-4)+1)
 50    DEF FNC(T)=(5-3*T)*(SGN(SGN(3*T-4)+1)
         -SGN(SGN(3*T-5)+1))
 60    DEF FND(T)=FNA(T)+FNB(T)+FNC(T)
 70    X=0:Y=0:T=M/(3 ^ N):K=0
 80    X=X+FND(T)/(2 ^ (K+1)):T=3*T:T=T-2*INT(T/2)
 90    Y=Y+FND(T)/(2 ^ (K+1)):K=K+1:IF K>INT(N/2)THEN 110
100    T=3*T:T=T-2*INT(T/2):GOTO 80
110    X=X+(M-2*INT(M/2))/(2 ^ K)):Y=Y+(M-2*INT(M/2))/(2 ^ K)
120    LINE -(480*X,198-198*Y), 1
130    M=M+1:IF M>3 ^ N THEN 150
140    GOTO 70
150    IF INKEY$= '' '' THEN 150
160    END
```

This program places the origin at the left margin, 199 pixels down, at (0,198). The generating function p from (7.1.1) is defined in four installments in lines 30–60 by making use of the unit step function

$$u(t,a) = \operatorname{sgn}(\operatorname{sgn}(t-a)+1).$$

In 80 and 100, t is reduced modulo 2 by means of $t = t - 2 * \operatorname{int}(t/2)$. Potential vertices occur at $t = m/3^n$, $m = 0, 1, 2, \ldots, 3^n$. x, y are evaluated by formula (7.3.1) for each such value of t in lines 80–110, and, in line 120, each vertex is joined to its predecessor by a straight line. For $n = 7$ and beyond, the lines become so dense that it becomes increasingly difficult to discern the approximating polygon.

The following program is obtained from Program 4 by some simple modifications. The evaluation of x and y in lines 90 and 100 is the same as in lines 80 and 90 of Program 4 but now the remainder term is not added on as it was in line 110 of Program 4.

Program 5
(generating the graphs of the a_n approximations)

```
10    DEFDBL T,X,Y
20    CLS:PRINT:Enter n = '';:INPUT N:SCREEN 2:
        PSET (0,198):M=1
30    P=2*N
40    DEF FNA(T)=(3*T-1)*(SGN(SGN(3*T-1)+1)
        -SGN(SGN(3*T-2)+1))
50    DEF FNB(T)=SGN(SGN(3*T-2)+1)-SGN(SGN(3*T-4)+1)
60    DEF FNC(T)=(5-3*T)*(SGN(SGN(3*T-4)+1)
        -SGN(SGN(3*T-5)+1))
70    DEF FND(T)=FNA(T)+FNB(T)+FNC(T)
80    X=0:Y=0:T=M/(3 ^ P):K=0
90    X=X+FND(T)/(2 ^ (K+1)):T=3*T:T=T-2*INT(T/2)
100   Y=Y+FND(T)/(2 ^ (K+1):K=K+1:IF K>N-1 THEN 120
110   T=3*T:T=T-2*INT(T/2):GOTO 90
120   LINE -(480*X,198-198*Y),1
130   M=M+1:IF M>3 ^ P THEN 150
140   GOTO 80
150   IF INKEY$='' '' THEN 150
160   END
```

A.2. Theorems from Analysis

A.2.1. Binary and Other Representations

The digits 0, 1, are called *binary digits*. If $t \in (0,1)$, then there are, for every $n = 1, 2, 3, \ldots$ binary digits $b_1, b_2, b_3, \ldots$ such that

$$b_1/2+b_2/2^2+b_3/2^3+\cdots+b_n/2^n \leq t < b_1/2+b_2/2^2+b_3/2^3+\cdots+b_n/2^n+1/2^n$$

and any such number has at least one and, at most, two binary representations

$$t = b_1/2 + b_2/2^2 + b_3/2^3 + \cdots = 0_2 b_1 b_2 b_3 \ldots .$$

If a binary representation is finite, $t = 0_{\dot{2}}b_1b_2b_3 \ldots b_n$, $b_n = 1$, then it is equivalent to the infinite binary representation $0_{\dot{2}}b_1b_2b_3 \ldots (b_n - 1)\bar{1}$. The number 1 has the binary representation $1 = 0_{\dot{2}}\bar{1}$. Analogous results hold for ternary, quaternary, etc.... representations (Randolph [1], pp. 29–34).

A.2.2. Condition for Non-Differentiability

Let $f : \mathcal{I} \to \mathbb{R}$. If there are two sequences $\{a_n\}, \{b_n\}$ with $0 < a_n \leq t < b_n < 1$, and $\{a_n\} \to t$, $\{b_n\} \to t$, such that

$$\lim_{n \to \infty} (f(a_n) - f(b_n))/(a_n - b_n)$$

does not exist, then f is *not* differentiable at t (Randolph [1], p. 365).

A.2.3. Completeness of the Euclidean Space

If $\{a_n\}$, with $a_n \in \mathbb{E}^m$, is a *Cauchy sequence*, i.e., for every $\varepsilon > 0$ there is a $N(\varepsilon)$ such that $||a_j - a_k|| < \varepsilon$ for all $j, k > N(\varepsilon)$, then there is an $a \in \mathbb{E}^m$ such that $\lim_{k \to \infty} a_k = a$. A space with this property is called complete (Apostol [1], p. 66, Bartle [1], p. 117, Randolph [1], p. 132, and Sagan [1], p. 223).

A.2.4. Uniform Convergence

If the functions $f_n : \mathcal{I} \to \mathbb{R}^m, n = 1, 2, 3, \ldots$ are continuous and $\{f_n\}$ converges *uniformly* to $f : \mathcal{I} \to \mathbb{R}^m$, i.e., for every $\varepsilon > 0$, there is a $N(\varepsilon)$, independent of t, such that $||f_n - f|| < \varepsilon$ for all $n > N(\varepsilon)$, then f is continuous in $\mathcal{I}$ (Apostol [1], p. 394, Randolph [1], p. 309, and Sagan [1], p. 275).

A.2.5. Measure of the Intersection of a Decreasing Sequence of Sets

If $\mathcal{A}_1, \mathcal{A}_2, \mathcal{A}_3, \ldots$ are Lebesgue measurable and $\mathcal{A}_1 \supset \mathcal{A}_2 \supset \mathcal{A}_3 \supset \ldots$ and, for at least one n, $\Lambda(\mathcal{A}_n) < \infty$, then

$$\Lambda\left(\bigcap_{n=1}^{\infty} \mathcal{A}_n\right) = \lim_{n \to \infty} \Lambda(\mathcal{A}_n)$$

(Randolph [1], p. 190).

A.2.6. Cantor's Intersection Theorem

If $\mathcal{C}_1 \supset \mathcal{C}_2 \supset \mathcal{C}_3 \supset \ldots$, where the $\mathcal{C}_n$ are non-empty, compact subsets of $\mathbb{E}^n$, then $\bigcap_{n=1}^{\infty} \mathcal{C}_n$ is closed and non-empty. If diameter $(\mathcal{C}_n) \to 0$ as $n \to \infty$, then $\bigcap_{n=1}^{\infty} \mathcal{C}_n$ contains exactly one point (Apostol [1], p. 56, Bartle [1], p. 88).

A.2.7. Infinite Products

If $0 \leq u_n < 1$, then $\prod_{n=1}^{\infty} (1 - u_n) > 0$ if and only if $\sum_{n=1}^{\infty} u_n$ converges (Apostol [1], p. 209).

References

Abend, K., Harley, T.J., Kanal, L.N.

[1] "Classification of binary random patterns," *IEEE Trans. Inform. Theory IT-11*, 538–544 (1965).

Alexanderson, G.L.

[1] "George Pólya: A Biographical Sketch," *The Pólya Picture Album*, Birkhäuser, Basel, 1987.

Alexandroff, P.

[1] "Über stetige Abbildungen kompakter Räume," *Mathem. Ann. 96*, 555–571 (1927).

Alsina, J.

[1] "The Peano Curve of Schoenberg is Nowhere Differentiable," *J. Approx. Theory, 33*, 28–42 (1981).

Apostol, T.M.

[1] *Mathematical Analysis*, 2nd edition, Addison-Wesley Publishing Company, Reading, MA, 1974.

Bagemihl, F., Piranian, G.

[1] "Absolutely Convergent Power Series," *Ann. Univ. Sci. Budapest. Eötvös (Sect. Math.) III–IV*, 27–34 (1960/1961).

Barnsley, M.

[1] *Fractals Everywhere*, Academic Press, Boston, 1988.

Bartholdi, J.J.III, Platzman, L.K.

[1] "On O(n log n) planar travelling salesman heuristic based on spacefilling curves," *Oper. Res. Lett. 1*, 121–125 (1982).

[2] "A fast heuristic based on spacefilling curves for minimum-weight matching in the plane," *Inf. Proc. Lett. 17*, 177–180 (1983).

[3] "Heuristics based on spacefilling Curves for combinatorial problems in euclidean space," *Management Science 34, No. 3 (March 1988)*, 291–305.

[4] "Spacefilling Curves and the Planar Travelling Salesman Problem," *J.A.C.M. 36, No. 4 (October 1989)*, 719–737.

Bartle, R.G.

[1] *The Elements of Real Analysis*, John Wiley & Sons, New York, 1967.

Beck, A., Bleicher, M.N., Crowe, D.W.

[1] *Excursions into Mathematics*, Worth Publishers, Inc., New York, 1969.

Bertsimas, D., Grigni, M.

[1] "Worst-Case Examples for the Spacefilling Curve Heuristic for the Euclidean Travelling Salesman Problem" *Oper. Res. Lett. 8*, 241–244 (1989).

Bially, Th.

[1] "Space-Filling Curves: Their Generation and Their Application to Bandwidth Reduction,"*IEEE Trans. Inform. Theory IT-15, No. 6 (November 1969)*, 658–664.

Biermann, Kurt, R.

[1] Eugen Netto, *Dictionary of Scientific Biography, Vol. 10*, Charles Scribner's Sons, New York, 1970, p. 24.

Blaschke, W.
[1] "Kreis und Kugel," *Jahresber. d. DMV 24*, 195–207 (1914).
[2] *Kreis und Kugel*, 2nd ed., Walter de Gruyter & Co., Berlin, 1956 (The 1st edition appeared in 1916).

Bliss, G.A., Dickson, L.E.
[1] "Biographical Memoir of Eliakim Hastings Moore," National Academy of Sciences of the United States of America, Biographical Memoirs, XVII—fifth memoir, Washington, DC, 1936, pp. 83–99.

Blumenthal, L.M., Menger, K.
[1] *Studies in Geometry*, W.H. Freeman and Co., San Francisco, 1970.

Borel, É.
[1] *Éléments de la Théorie des Ensembles*, Éditions Albin Michel, Paris, 1949, pp. 279–286.

Bosznay, A.P.
[1] "A remark on a Paper of B.R. Gelbaum," *Z. Wahrsch. verw. Gebiete, 36*, 353–355 (1978).

Bryc, W.
[1] "A remark on continuous independent functions," *Ann. Univ. Sci. Budapest. Eötvös (Sect. Math.) 26*, 13–16 (1983).

Bumby, R.T.
[1] "The Differentiability of Pólya's Function," *Adv. in Mathem.*, *18*, 243–244 (1975).

Butz, A.R.
[1] "Space Filling Curves and Mathematical Programming " *Inf. and Control 12*, 314–330 (1968).
[2] "Convergence with Hilbert's Space-Filling Curve," *J. Comp. and Syst. Sc. 3*, 128–146 (1969).
[3] "Alternative Algorithm for Hilbert's Space-Filling Curve," *IEEE Trans. Comp. (April 1971)*, 424–426.

Calinger, R.G.
[1] Eliakim Hastings Moore, *Dictionary of Scientific Biography, Vol. 9*, Charles Scribner's Sons, New York, 1970, pp. 501–503.

Cantor, G.
[1] "Ein Beitrag zur Mannigfaltigkeitslehre," *Crelle J.*, *84*, 242–258 (1878).
[2] "Über unendliche lineare Punktmannigfaltigkeiten," *Math. Ann.*, *21*, 445–591 (1883).

Carathéodory, C.
[1] *Theory of Functions of a Complex Variable*, Chelsea Publishing Co., New York, 1964, p. 190.

Cesàro, E.
[1] "Sur la représentation analytique des régions, et des courbes qui les remplissent," *Bull. des Sc. Math., (2), 21*, 257–266 (1897).
[2] "Remarques sur la courbe de von Koch," *Atti della R. Academia delle Scienze fisiche e mathematiche, Napoli, Séries II, Vol. XII, No. 15*, 1–12 (1905).
[3] "Fonctions continues sans dérivées," *Archiv für Mathematik und Physik, Serie 3, Bd. X*, 57–63 (1906).

Cichoń, J., Morayne, M.
[1] "On Differentiability of Peano Type Functions III," *Proc. Am. Math. Soc., 92*, 432–438 (1984).

Davis, C., and Knuth, D.E.
[1] "Number representations and Dragon Curves I," *J. Recreational Mathem., 3*, 66–81 (1970).
[2] "Number representations and Dragon Curves II," *J. Recreational Mathem., 3*, 133–149 (1970).

Debski, W., Mioduszewski, J.
[1] "Simple plane images of the Sierpiński triangular Curve are nowhere dense," *Colloquium Mathematicum LIX*, 125–139 (1990).

Denjoy, A.
[1] "Sur la vie et l'oeuvre de Henri Lebesgue," *Institut de France, Academie des Sciences*, 1–30 (1946) (also in Lebesgue [3] pp. 35–65).
Denjoy, A., Felix, L., Montel, P.
[1] "Henri Lebesgue: Le savant, le professeur, l'homme," *L'Enseignement mathém.*, t. III, fasc. 1, 1–18 (also in Lebesgue [3], pp. 67–84).
Devinatz, A.
[1] *Advanced Calculus*, Holt, Rinehart, Winston, New York, 1968, p. 253.
Dinghas, A.
[1] "Zur Peano Abbildung einer Strecke," *Israel J. of Math.*, *7*, 211–216 (1969).
Donoghue, W.F., Jr.
[1] "Continuous Function Spaces Isometric to Hilbert Space," *Proc. Am. Math. Soc.*, *8*, 1–2 (1957).
Edgar, G.A.
[1] *Measure, Topology, and Fractal Geometry*, Springer-Verlag, New York, 1990.
Engelking, R.
[1] *General Topology*, Polska Academia Nauk, Instytut Matematyczny, Monografie Matematyczne, Vol. 60, Warszawa, 1977.
Fréchet, M.
[1] "Sur quelques points du Calcul Fonctionell," *Rend. Palermo 27*, 1–74 (1906).
Freudenthal, H.
[1] David Hilbert, *Dictionary of Scientific Biography, Vol. 6*, Charles Scribner's Sons, New York, 1970, pp. 388–394.
[2] Konrad Knopp, *Dictionary of Scientific Biography, Vol. 7*, Charles Scribner's Sons, New York, 1970, pp. 411–412.
Gardner, M.
[1] "Mathematical Games," *Scientific American, 216 (March 1967)*, 124–125; *216 (April 1967)*, 118–120; *217 (July 1967)*, 115.
[2] *The unexpected Hanging and Other Mathematical Diversions*, Simon and Schuster, New York, 1969, Chapter 19.
Garsia, A.M.
[1] "Combinatorial Inequalities and Smoothness of Functions," *Bull. Am. Math. Soc. 82*, 157–170 (1976).
Gelbaum, B.R. and Olmsted J.M.H.
[1] *Counterexamples in Analysis*, Holden Day, Inc., San Francisco, 1965.
Gelbaum, B.R.
[1] "Independence of Events and of Random Variables," *Z. Wahrsch. verw. Gebiete*, *36*, 333–343 (1976).
Golomb, S.W.
[1] "Replicating Figures on the Plane," *Math. Gaz.*, *48*, 403–412 (1964).
Grattan-Guiness, I.
[1] "From Weierstrass to Russell: A Peano Medley," Celebrazioni in memoria di Giuseppe Peano nel cinquantenario della morte, Atti del Convegno organizzato dal Dipartimento di Matematica dell'Università di Torino, 27–28 ottobre 1982, Torino, 1986.
Greenberg, M.
[1] *Lectures on Algebraic Topology*, W.A. Benjamin, New York, 1967, p. 82.
Hahn, H.
[1] "Über die Abbildung einer Strecke auf ein Quadrat," *Annali di Matematica, Serie III, XXI*, 33–55 (1913).
[2] "Über die allgemeinste ebene Punktmenge, die stetiges Bild einer Strecke ist," *J. Ber. d. DMV*, *23*, 318–322 (1914).
[3] "Mengentheoretische Charakterisierung der stetigen Kurven," *S.-B. Kaiserl. Akad. Wiss. Wien, Abt. 2a (Math.) 123*, 2433–2487 (1914).
[4] "Über die stetigen Kurven der Ebene," *Mathem. Zeitschrift*, *9*, 66–73 (1921).

[5] "Über die Komponenten offener Mengen," *Fund. Math. 2*, 189–192 (1921).
[6] "Über irreduzible Kontinua," *S.-Ber. Öst. Akad. Wiss. Wien, Abt. 2a* (Math) *130*, 217–250 (1921).
[7] "Über stetige Streckenbilder," Atti del Congresso Internazionale dei Mathematici, Bologna, 3–10 September 1928, *VI*, 217–220.
[8] "Über die Anordnungssätze der Geometrie," *Monatsh. f. Math. u. Physik*, *19*, 289–303 (1908).

Hausdorff, F.
[1] *Set Theory*, Chelsea, New York, 1962.

Hawkins, T.
[1] Henri Léon Lebesgue, *Dictionary of Scientific Biography, Vol. 8*, Charles Scribner's Sons, New York, 1970, pp. 110–112.

Hilbert, D.
[1] "Über die stetige Abbildung einer Linie auf ein Flächenstück," *Math. Annln.*, *38*, 459–460 (1891).
[2] *The Foundations of Geometry*, Open Court Publishing Company, La Salle, IL, 1938.

Hiriart-Urruti, J.-B.
[1] "Une courbe étrange venue d'ailleurs," *Revue de Mathematiques speciales (1986)*, 229–230.

Hlawka, E.
[1] "Über eine Klasse von Näherungspolygonen zur Peanokurve," *J. Number Theory*, *43*, 93–108 (1993).

Holbrook, J.R.
[1] "Stochastic Independence and Space-filling curves," *Am. Math. Monthly*, *88*, 426–432 (1981).

Hurwitz, A.
[1] "Über die Entwicklung der Allgemeinen Theorie der analytischen Funktionen in neuerer Zeit," Verhandlungen des ersten internationalen Mathematiker-Kongresses in Zürich vom 9. bis 11. August 1897, B.G. Teubner, Leipzig, 1898.

Jessen, B.
[1] "Über eine Lebesgue'sche Integrationstheorie für Funktionen unendlich vieler Veränderlicher," C.R. du septième Congrès des mathematiciens scandinaves tenu à Oslo, 19–22 août 1929, 34–36.

Jordan, C.
[1] *Cours d'Analyse de l'École Polytechnique*, Tome Premier, Troisiéme Édition, Gauthier-Villars, Paris, 1959.
[2] *l'Intermediaire des Mathematiciens*, Vol. 1, Gauthier-Villars, Paris, 1894, p. 23.

Kahane, J.-P.
[1] "Courbes étranges, ensembles minces," *Bull. de l'Asociation des Professeurs de Mathematiques (A.P.M.)*, *275/276*, 325–339 (1970).

Kamke, E., Zeller, K.
[1] "Konrad Knopp †," *Jahresber. d. DMV*, *60*, 44–49 (1957).

Kennedy, H.C.
[1] Giuseppe Peano, *Dictionary of Scientific Biography, Vol. 10*, Charles Scribner's Sons, New York, 1970, pp. 441–444.

Knaster, B.
[1] Stefan Mazurkiewicz, *Dictionary of Scientific Biography, Vol. 9*, Charles Scribner's Sons, New York, 1970, pp. 248–250.

Knopp, K.
[1] "Einheitliche Erzeugung und Darstellung der Kurven von Peano, Osgood und von Koch," *Arch. Math. Phys.*, *26*, 103–115 (1917).

Koch, H. von

[1] "Sur une courbe continue sans tangente obtenue par une construction géométrique élémentaire," *Arkiv for Matematik och Fysik, 1*, 681–702 (1903–1904).

[2] "Une méthode géométrique élémentaire pour l'étude de certaines questions de la théorie des courbes planes," *Acta Mathematica, 29–30*, 145–174 (1905–1906).

Kowalewski, G.

[1] "Über Bolzano's nichtdifferenzierbare stetige Funktion," *Acta Math., 44*, 315–319 (1923).

Kuratowski, K.

[1] *Topology*, Vol. I, new edition, revised and augmented, Academic Press, New York, 1966.

[2] Waclaw Sierpiński, *Dictionary of Scientific Biography, Vol. 12*, Charles Scribner's Sons, New York, 1970, pp. 426–427.

[3] *A Half Century of Polish Mathematics*, Pergamon Press, Oxford, 1980.

Kuttler, J.R.

[1] "Nowhere-Differentiable Functions," *Am. Math. Monthly, 99*, 565–566 (1992).

Lance, T. and Thomas, E.

[1] "Arcs with Positive Measure and a Space-filling Curve," *Amer. Math. Monthly, 98*, 124–127 (1991).

Langford, C.D.

[1] "Uses of a geometric Puzzle," *Math. Gaz., XXIV*, 209–211 (1940).

Lax, P.D.

[1] "The Differentiability of Pólya's Function," *Adv. in Mathem., 10, No. 3*, 456–464 (1973).

Lebesgue, H.

[1] *Leçons sur l'Intégration et la Recherche des Fonctions Primitives*, Gauthier-Villars, Paris, 1904, pp. 44–45.

[2] "Sur les fonctions représentables analytiquement," *J. de Math., 6(1)*, 139–216 (1905).

[3] *Oeuvres Scientifiques*, Vol. I, L'Enseignement Mathématique, Institut de Mathématiques, Université de Genève, 1972.

Mandelbrot, B.B.

[1] *Fractals, Form, Chance, and Dimension*, W.H. Freeman and Company, San Francisco, 1977.

Mayrhofer, K.

[1] "Hans Hahn †," *Monatsh. f. Mathem. u. Physik, 41*, 221–238 (1934).

Mazurkiewicz, St.

[1] "O arytmetyzacji kontinuow," *C.R. Soc. Sc. Varsovie, VI*, 305–311 (1913) (also in Mazurkiewicz [6], 37–41, under the title "Sur l'arithmetisation des continus").

[2] "O arytmetyzacji kontinuow *II*," *C. R. Soc. Sc. Varsovie, VI*, 941–945 (1913) (also in Mazurkiewicz [6], 42–45, under the title "Sur l'arithmetisation des continus II").

[3] "Sur les lignes de Jordan", *Fund. Math., 1*, 166–209 (1913) (also in Mazurkiewicz [6], 76–113).

[4] "O punktach wielokrotnych krzywych wypelniajacych obszar plaski" ("Sur les points multiples des courbes qui remplissent une aire plane") *Prace Matematyczno-Fizyczne, XXV*, 113–120 (1914).

[5] "Sur une classification des points situés sur un continue arbitraire," *C.R. Soc. Sc. Varsovie, IX*, 441, 442 (1916).

[6] *Travaux de Topolgie et ses Applications*, PWN- Polish Scientific Publishers, Warszawa, 1969.

Meschkowski, H.

[1] George Cantor, *Dictionary of Scientific Biography, Vol. 3*, Charles Scribner's Sons, New York, 1970, pp. 52–58.

Milne, St.C.
[1] "Peano Curves and Smoothness of Functions," *Adv. in Math.*, *35*, 129–157 (1980).
Mioduszewski, J.
[1] "Odwzorowania Peani, czyli o rozcinaniu a potem sklejaniu kwadratów i trójkatów" *Delta Popularny Miesiecznik Matematyczno-Fizyczny*, *7*, 1–4 (1977).
[2] "Peano Maps through Cuttings," *Center for Mathematical Culture, Special Issue (1990)*, 22–23.
Montel, P.
[1] "Notice nécrologique sur M. Henri Lebesgue," *C.R.T. 213, No. 5*, 197–200 (1942) (also in Lebesgue [3], pp. 31–34).
Moore, E.H.
[1] "On certain crinkly curves," *Trans. Amer. Math. Soc.*, *1*, 72–90 (1900).
Morayne, M.
[1] "On Differentiability of Peano Type Functions," *Colloquium Mathematicum*, *LIII*, 129–132 (1987).
[2] "On Differentiability of Peano Type Functions II," *Colloquium Mathematicum*, *LIII*, 133–135 (1987).
Netto, E.
[1] "Beitrag zur Mannigfaltigkeitslehre," *Crelle J.*, *86*, 263–268 (1879).
Ohno, O. and Ohiyama K.
[1] "A catalog of symmetric self-similar space-filling curves," *J. Recreational Mathem.*, *23*, 161–173 (1991).
[2] "A catalog of non-symmetric self-similar space-filling curves," *J. Recreational Mathem.*, *23*, 247–254 (1991).
Olmsted, J.M.H.
[1] *Real Variables*, Appleton-Century-Crofts, Inc., New York, 1959, p. 342.
Orman Quine, W. van
[1] "Peano as Logician," Celebrazioni in memoria di Giuseppe Peano nel cinquantenario della morte, Atti del Convegno organizzato dal Dipartimento di Matematica dell'Università di Torino, 27–28 ottobre 1982, Torino, 1986.
Osgood, W.F.
[1] "On the Existence of the Green's Function for the most general simply connected plane region," *Trans. Am. Math. Soc.*, *1*, 310 (1900).
[2] "A Jordan Curve of positive Area," *Trans. Amer. Math. Soc.*, *4*, 107–112 (1903).
[3] *Funktionentheorie*, 1. Bd., Chelsea Publ. Co., New York, 1963, p. 171.
Patrick, E.A., Anderson, D.R., Bechtel, F.K.
[1] "Mapping Multidimensional Space to One Dimension for Computer Output Display," *IEEE Trans. Comp. C-17, No. 10 (October 1968)*, 949–953.
Peano, G.
[1] "Sur une courbe qui remplit toute une aire plane," *Math. Annln.*, *36*, 157–160 (1890).
[2] *l'Intermediaire des Mathematiciens, Vol. 3*, Gauthier-Villars, Paris, 1896, p. 39.
Peitgen, H.-O., Richter, P.H.
[1] *The Beauty of Fractals*, Springer-Verlag, New York, 1986.
Peitgen, H.-O., Saupe, D. (Editors)
[1] *The Science of Fractal Images*, Springer-Verlag, New York, 1988.
Piranian, G., Titus, C.J., Young, G.S.
[1] "Conformal Mappings and Peano Curves," *Michigan Math. J.*, *1*, 69–72 (1952).
Pólya, G.
[1] "Über eine Peanosche Kurve," *Bull. Acad. Sci. Cracovie (Sci. math. et nat. Série A) (1913)*, 1–9.

Rahm, G. de
[1] "Un peu de mathématiques apropos d'une courbe plane," *Elem. Math.*, *2*, 73–76, 89–97 (1947).
Randolph, J.F.
[1] *Basic Real and Abstract Analysis*, Academic Press, New York, 1968, pp. 365–367.
Robertson, A.G.
[1] "Multiplicativity of the Uniform Norm and Independent Functions," *Bull. Austral. Math. Soc.*, *42*, 153–155 (1990).
Rosenthal, A.
[1] "Über Peano Flächen und ihren Rand," *Mathem. Z.*, *10*, 102–104 (1921).
Rudin, W.
[1] *Principles of Mathematical Analysis*, First Ed., McGraw-Hill, New York, 1953.
Sagan, H.
[1] *Advanced Calculus*, Houghton-Mifflin Company, Boston, 1974.
[2] "Approximating Polygons for Lebesgue's and Schoenberg's Space-filling Curves," *Amer. Math. Monthly*, *93*, 361–368 (1986).
[3] "Some Reflections on the Emergence of Space-filling Curves: The Way it could have happened and should have happened, but did not happen," *Franklin J.*, *328*, 419–430 (1991).
[4] "An Elementary Proof that Schoenberg's Space-filling Curve is nowhere differentiable," *Math Mag.*, *65*, 125–128 (1992).
[5] "On the Geometrization of the Peano Curve and the Arithmetization of the Hilbert Curve," *Internat. J. Math. Ed. Sc. Tech*, *23*, 403–411 (1992).
[6] "Approximating Polygons for the Sierpiński-Knopp Curve," *Bull. Acad. Sci. Polonaise*, *40*, 19–29 (1992).
[7] "Nowhere Differentiability of Sierpiński's Space-filling Curve," *Bull. Acad. Sci. Polonaise*, *40*, 217–220 (1992).
[8] "A three-dimensional Hilbert-Curve," *Internat. J. Math. Ed. Sc. Tech.*, *24*, 541–545 (1993).
[9] "An analytic Proof of the Nowhere Differentiability of Hilbert's Space-filling Curve," *Franklin J.*, *330*, 763–766 (1993).
[10] "A Geometrization of Lebesgue's Space-filling Curve," *Math. Intelligencer*, *15*, No. 4, 37–43 (1993).
[11] "The Coordinate Functions of Sierpiński's Space-Filling Curve are Nowhere Differentiable," *Bull. Acad. Sci. Polonaise*, *41*, 73–75 (1993).
[12] "The taming of a Monster: A parametrization of the von Koch curve," *Internat. J. Math. Ed. Sc. Tech.*, to appear.
Salem, R., Zygmund, A.
[1] "Lacunary Power Series and Peano Curves," *Duke J.*, *12*, 569–578 (1945).
Schaeffer, A.C.
[1] "Power Series and Peano Curves," *Duke J.*, *21*, 383–390 (1954).
Schoenberg, I.J.
[1] "The Peano-Curve of Lebesgue," *Bull. Amer. Math. Soc.*, *44*, 519 (1938).
[2] Mathematical Time Exposures, *Math. Assoc. Amer.*, *1982*, 135–148.
[3] Personal letter to the author, 7 May 1986.
Schoenflies, A.
[1] *Die Entwicklung der Lehre von den Punktmannigfaltigkeiten*, 2. Teil, B.G. Teubner, Leipzig, 1908.
Sierpiński, W.
[1] "Sur une nouvelle courbe continue qui remplit toute une aire plane," *Bull. Acad. Sci. de Cracovie (Sci. math. et nat., Série A) (1912)*, 462–478 (also in Sierpiński [11], pp. 52–66).
[2] "O krzywych, wypolniajacych kwadrat," *Prace matematyczno-fizyczne*, *XXIII*, 193–219 (1912).
[3] "Sur une courbe non quarrable," *Bull. Acad. Sci. de Cracovie (Sci. math. et nat., Série A) (1913)*, 254–263.

[4] "O krzywej, której każdy punkt jest punktem rozgalezienia," *Prace Mat.-Fiz. 27*, 77–86 (1916) (also in Sierpiński [11], 99–106 under the title Sur une courbe dont tout point est un point de ramification").

[5] "Sur une courbe dont tout point est un point de ramification," *C.R. Acad. Sci., Paris, 160*, 302–305 (1915).

[6] "O krivoy, soderzhashey v sebe obraz vsyakoy krivoy," *Matematichesky sbornik, 30*, 267–287 (1916) (also in Sierpiński [11], 107–119, under the title "Sur une courbe cantorienne qui contient une image biunivoque et continue de toute courbe donnée").

[7] "Sur une courbe cantorienne qui contient une image biuivoque et continue de toute courbe donnée," *C.R. de l'Academie des Sciences Paris, 172*, 629–632 (1916).

[8] "Sur une condition pour qu'un continu soit une courbe jordanienne," *Fund. Math., 1*, 44–60 (1920) (also in Sierpiński [11], 308–321).

[9] "Remarque sur la courbe péanienne," *Wiadomości Matematyczne, XLII*, 1–3 (1937) (also in Sierpiński [12], 369–371).

[10] *Cardinal and Ordinal Numbers*, 2nd edition, Éditions Scientifiques de Pologne, Warszawa, 1965.

[11] *Oeuvres Choisies*, Tome II, Éditions Scientifiques de Pologne, Warszawa, 1975.

[12] *Oeuvres Choisies*, Tome III, Éditions Scientifiques de Pologne, Warszawa, 1976.

Stachel, H.

[1] Laudatio, *Nouvelles Mathematiques Internationales, No. 154, Österreichische Mathematische Gesellschaft, Wien, (August 1990)* 4–9.

Steinhaus, H.

[1] "La courbe de Peano et les fonctions indépendantes," *C.R. Acad. Sci., Paris, 202*, 1961–1963 (1936).

Steinitz, E.

[1] "Stetigkeit und Differentialquotient," *Math. Ann., 52*, 59–69 (1899).

Stromberg, K., Tseng, Shiojenn

[1] "Simple plane Arcs of positive Area," *Expo. Math. 12* (1994).

Strubecker, K.

[1] *Einführung in die höhere Mathematik*, Bd. II, R. Oldenbourg, München-Wien, 1967.

Takagi, T.

[1] "A simple example of a continuous function without derived function," *J. Phys. School. Tokyo, 14*, 1–2 (1904).

Torhorst, M.

[1] "Über den Rand der einfach zusammenhängenden ebenen Gebiete," *Mathem. Zeitschrift, 9*, 45–65 (1921).

Veblen, O.

[1] A system of Axioms for geometry, *Trans. Am. Math. Soc., 5*, 343–384 (1904).

[2] Theory of plane curves in non-metrical analysis situs, *Trans. Am. Math. Soc., 6*, 83–98 (1905).

Walsh, J.L.

[1] William Fogg Osgood, *Dictionary of Scientific Biography, Vol. 10*, Charles Scribner's Sons, New York, 1970, pp. 244–245.

Willard, St.

[1] *General Topology*, Addison-Wesley, Reading, MA., 1970, pp. 219–222.

Wirth, N.

[1] *Algorithms + Data Structures = Programs*, Prentice-Hall, Inc., Englewood Cliffs, NJ, 1976.

Wunderlich, W.

[1] "Irregular Curves and Functional Equations," Ganita (Proc. Benares Math. Soc.) *5*, 215–230 (1954).

[2] "Una generazione comune di diverse curve patologiche," *Atti VI Congr. U.M.I. Napoli*, 426–427 (1959).

[3] "Über Peano-Kurven," *Elem. Math.*, *28*, 1–10 (1973).

Yost, D.

[1] "Space-filling curves and universal normed spaces," *Ann. Univ. Sci. Budapest. Eötvös (Sect. Math.) 27*, 33–36 (1984).

Index

Abend, K., 30, 177
accumulation point, 88
accumulation points of the
  Cantor set, 73
$\aleph_0$-dimensional
  Schoenberg curve, 128
  space-filling curve, 112
Alexanderson, G.L., IX, 62, 177
Alexandroff, P., 100, 177
Alsina, J., 121, 125, 177
analytic function
  representation of a space-filling curve
    by an, 112 ff.
Anderson, D.R., 30, 182
Antosik, Piotr, IX
Apostol, T.M., 174, 175, 177
approximating polygons to the
  Heighway dragon, 163
  Hilbert curve, 21, 22
  Lebesgue curve, 79 ff.
  Peano curve, 42
  Pólya curve, 67
  Schoenberg curve, 123 ff.
    plotting program, 172, 173
  Sierpiński-Knopp curve, 60 ff.
arcwise connected, 117
attractor set, 150, 151, 154, 155

Bagemihl, F., 112, 177
Banach, St., 86
Barnsley, M., 149, 150, 156, 158, 177
Bartholdi III, J.J., 30, 177
Bartle, R.G., 174, 175, 177
Bechtel, F.K., 182
Beck, A., 136, 177
Bertsimas, D., 30, 177
Betsch, G., IX
Bially, Th., 30, 177
Biermann, K.R., 6, 177
bijective map, 3
binary representation, 3, 173
Birkhoff, C.D., 25
Blaschke, W., 152, 178
  selection theorem of, 152
Bleicher, M.N., 136, 177
Bliss, G., 25, 178
Blumenthal, L.M., 161, 162, 178
Bôcher, M., 133
Bolza, O., 25
Bolzano-Kowalewski curve, 149, 167
Boltzman, L., 86
Borel, É., 13, 76, 178
Bosznay, A.P., 109, 178
bounded set, 91
Bryc, W., 109, 178
Bumby, R.T., 62
Burckel, R.B., IX
Butz, A.R., 29, 30, 178

Calinger, R.G., 25, 178
Cantor, Georg, 1, 69, 178
  biography, 5
Cantor
  brush, 117
  curve, 159 ff.
  dust, 145
  function, 74–76
  set, 3, 69, 145
    as invariant attractor set, 155
    as perfect set, 73
    cardinality of, 69
    measure of, 74
    of positive measure, 82
    similarity dimension of, 158
    ternary representation of, 72
Cantor's intersection theorem, 175
Carathéodory, C., 150, 178
cartesian space, 2
Cauchy sequences, 174
  of non-empty compact subsets of a
    complete metric space,
    151 ff.
Cesàro, E., 40, 78, 146, 178

Cesàro's representation of the Peano curve, 40 ff.
Chandler, R.E., IX
Cichoń, J., 79, 178
closed
  set, 88
  unit cube, 2
  unit interval, 2
  unit square, 2
closure, 88
Cole, F.N., 133
compact set, 91
  as continuous image of the Cantor set, 100
  that is pathwise connected but not locally connected, 103
complement of a set, 3
completeness of $\mathbb{E}^n$, 174
completeness of space of non-empty compact subsets of complete metric space, 151 ff.
complex numbers, field of, 2
complex representation of Hilbert curve, 13 ff.
computer programs, 169–173
  computation of the nodal points of the
    Hilbert curve, 169
    Peano curve, 170
    Sierpiński-Knopp curve, 171
  plotting programs for approximating polygons of the Schoenberg curve, 172, 173
connected set, 94 ff.
continuity, 85 ff.
  global characterization of, 89, 90
continuity of the
  Hilbert curve, 12
  Lebesgue curve, 77, 79
  Peano curve, 33
  Schoenberg curve, 120
  Sierpiński-Knopp curve, 52
continuous extension from a dense set, 93
continuous image of
  a compact locally connected set, 99
  a compact set, 92
  a connected set, 95
  an interval, 93
continuum, 160
contraction mapping theorem, 151, 154
convergence
  uniform, 174
Courant, R., 62
Crowe, D.W., 136, 177
curve, 4
  Bolzano-Kowalewski, 149, 167
  Cantor, 161
  Hilbert, 9 ff.
  Jordan, 6
  Kieswetter, 166
  von Koch, 54, 78, 140, 146, 149, 156, 165
  Osgood, 136
  parameter representation of, 5
  Peano, 31 ff.
  Rham, 149
  Schoenberg, 119 ff.
  Sierpiński, 49 ff.
  Sierpiński-Knopp, 51 ff.
  space-filling, 5
  Takagi, 166

Davis, C., 162, 165, 178
devil's staircase, 75, 76
  generation by iterated function system, 166
Debski, W., 178
Denjoy, A., 76, 77, 179
dense, 88
Devinatz, A., 179
Dick, A., IX
Dickson, L.E., 25, 178
diffeomorphism, 4, 10
difference set, 82
dilation of a set, 150
dimension
  Hausdorff, 158
  similarity, 156 ff.
  topological, 156
Dinghas, A., 112, 179
direct image, 4, 88
disconnected set, 94
domain, 4
Donoghue, W.F., Jr., 109, 179
dyadic
  discontinuum, 116
  representation, 4
  set, 116

Eckhart, L., 43
Edgar, G.A., 149, 158, 162, 179
eneadic, 7, 37
Engelking, R., 108, 179
Eötvös, L., 62
"equal by definition" symbol, 4
Escherich, Gustav, Ritter von, 86
η-chain, 96
euclidean
  norm, 3
  space, 3
    completeness of, 174
Euler-Knopp process, 56
Euler spline, 121
excluded middle thirds, set of, 69

Faulkner, Gary D., IX
Fejér, L., 62
Fekete, M., 62
Felix, L., 76, 77
finite subcover, 100
four-dimensional space-filling curve, 110
fractals, 23, 145 ff.
  examples of, 145 ff.
  literature on, 149
  self-similar, 149
Frank, W., IX
Fréchet, M., 108, 179
  space, 108
Freudenthal, H., 10, 179
Fricke, R.R., 149
Furtwängler, Ph., 43

Gardner, M., 162, 179
Garsia, A.M., 109, 111, 179
Gelbaum, B.R., 109, 133, 140, 179
geometric generation of the
  Hilbert curve, 9 ff.
  Peano curve, 34 ff.
  Sierpinski-Knopp curve, 51 ff.
Giacardi, L.M., IX
global continuity, 89, 90
Gödel, K., 10
Golomb, S.W., 30, 179
Grattan-Guiness, I., 31, 179
Greenberg, M., 6, 98, 179
Grigni, M., 30, 177
Gruber, P.M., IX

Hadamard, J., 62
Hahn, H., 2, 67, 85, 98–101, 106–108, 131, 179
  biography, 86
Hahn-Mazurkiewicz theorem, 106 ff.
Hardy, G.H., 62
Harley, T.J., 30, 179
Hausdorff, F., 85, 98, 99, 106, 180
  dimension, 158
  distance, 150, 151
  space, 108
  theorem, 99 ff.
Hawkins, T., 77, 180
Heine-Borel
  property, 91
  theorem, 91
Heighway, J.E., 162
Heighway dragon, 162 ff.
  approximating polygons for, 163
  as a space-filling curve, 164
  as one of the latest monsters, 165
  similarity dimension of, 164
Hestenes, M., 25
Hilbert, D., VIII, 1, 62, 180
  biography, 9
Hilbert's space-filling curve, 9 ff., 148, 149
  approximating polygons for, 21
  arithmetization of, 18
  as invariant attractor set, 155
  as measure preserving map, 111
  complex representation, 13 ff.
  computation of nodal points, 169, 170
  continuity of, 12
  definition, 11
  Moore's version, 24
  nowhere differentiability, 12, 19 ff.
  similarity dimension, 159
  three-dimensional, 26 ff.
Hiriart-Urruty, J.-B., 121, 180
Hlawka, E., IX, 127, 180
Holbrook, J.R., 109, 180
homeomorphism, 4
Hurwitz, A., 9, 62, 132, 180

infinite products, 175
initial set, 23,
  independence of, 155
injective map, 3
interior of a set, 88
interior point, 88
intersection theorem, Cantor's, 175
interval as connected subset of $\mathbb{R}^n$, 95
invariant attractor set, 150, 151, 154 ff.
  existence and uniqueness, 155
inverse image, 88
Iséki, Kiyoshi, 128
iterated function system, 150

Jacobs, K., IX
Janiszewski, Z., 49, 86
Jessen, B., 112, 180
joins, 133
Jones, F.B., 132
Jordan, C., 131, 132, 180
Jordan
  content, 3
  content zero, 82
  curve, 6, 131
    definition, 132
    Osgood's, 132 ff.
  of positive two-dimensional Lebesgue measure, 131 ff., 135
Julia, G., 62

Kahane, J.-P., 180
Kamke, E., 56, 180

Kanal, L.N., 30, 177
Kennedy, H.C., 31, 180
Kieswetter curve, 166
Klein, F., 24, 62, 133, 149
Knaster, B., 86, 180
Knopp, K., 1, 54, 55, 136, 140, 146, 180
  biography, 55
Knopp's
  generation of von Koch curve, 140
  Osgood curves, 137 ff.
  representation of Sierpiński's space-filling curve, 51 ff.
Knuth, D.E., 162, 165, 178
Koch, H. von, 54, 181
  curve, 54, 78, 140, 146, 149, 156, 165
    generation by iterated function system, 146
    similarity dimension of, 158
Kowalewski, G., 149, 181
Krames, J., 43
Krasinkiewicz, J., IX
Kruppa, E., 43
Kronecker, L., 5, 6, 24
Kummer, E.E., 5
Kuratowski, K., 50, 86, 112, 181
Kuttler, J.R., 166, 181

lacunary power series, 2, 112
Lance, T., 140, 181
Landau, E., 44
Langford, C.D., 181
Lax, P.D., IX, 63, 181
Lebesgue, H.L., 1, 76, 181
  biography, 76
Lebesgue
  measure, 3
    of von Koch curve, 139, 140
    of Mazurkiewicz continuum, 161
    of modified Sierpiński triangle, 156
    of Sierpiński gasket (triangle), 156
    of Sierpiński sponge, 162
    of Sierpiński triangle (gasket), 156
    zero, 82
Lebesgue's space-filling curve, 76 ff.
  a.e. differentiability of, 78
  approximating polygons of, 79
  continuity of, 77, 81
Lehmer, D.N., 25
Leitmotiv, 23, 146
Lichtenstein, L., 56
Lindemann,. C.L.F., 9
line segment, continuous images of, 85 ff.
Littlewood, J.E., 62
local connectedness, 98 ff.
  uniform, 102
Loskot, K., IX, 111

Mandelbrot, B.B., 145, 149, 161, 183
mapping of Cantor set onto a square, 70
  continuity of, 70
  nowhere differentiability of, 70
matrices, notation, 3
Mayer, A., 86
Mayrhofer K., 43, 86, 181
Mazurkiewicz, St., 2, 6, 49, 67, 85, 100, 106, 108, 159, 161, 181
  biography, 87
Mazurkiewicz continuum, 160
McShane, E.J., 25
measure of
  Cantor set, 74
  intersection of decreasing sequence of sets, 174
measure preserving maps, 111
Menger, K., 161, 162, 178
Menger sponge, 161
Mertens, F., 86
Meschkowski, H., 5, 181
Milne, St. C., 109, 112, 182
Minkowski, H., 86
Mioduszewski, J., 30, 178, 182
modified Sierpiński triangle, 148
  similarity dimension of, 159
Montel, P., 76, 77, 182
Moore, E.H., 1, 10, 21, 24, 34, 35, 133, 182
  biography, 24
Moore's version of the Hilbert curve, 24
Moran's open set condition, 158
Morayne, M., 79, 178, 182
Mund, B., IX

n-dimensional space-filling curve, 112
neighborhood, 88
  deleted, 88
neighborhood of a compact non-empty subset of a complete metric space, 150
Netto, E., 1, 5, 6, 182
  biography, 6
Netto's theorem, 6
  proof of, 97
Newton, H.H., 24
nodal points, 21
  computation of, 169, 170, 171
non-differentiability
  condition for, 174
notation, 2 ff.
nowhere differentiability of
  Hilbert's space-filling curve, 12, 19
  Peano's space-filling curve, 34
  Schoenberg's space-filling curve, 121
  Sierpiński-Knopp space-filling curve, 58

octals, 3
Ohiyama, K., 182
Ohno, O., 182
Olmsted, J.M.H., 13, 133, 140, 179, 182
open set, 88
Orman Quine, W. van, 31, 182
Osgood, W.F., 1, 54, 131, 182
  biography, 133
Osgood curves, 54, 136
Osgood curves with
  Hilbert curve as limiting arc, 140
  Lebesgue curve as limiting arc, 142
  Peano curve as limiting arc, 134
  Sierpinski-Knopp curve as limiting arc, 137

parametrization of a curve, 5
pathwise connectedness, 101 ff.
Patrick, E.A., 30, 182
Pawlikowska-Brozek, Z., IX
Peano, G., VIII, 10, 46, 132, 182
  biography, 31
Peano's space-filling curve, 31 ff.
  approximating polygons, 42, 43
  as invariant attractor set, 155
  as measure preserving map, 111
  Cesàro's representation of, 40 ff.
  computation of nodal points, 170, 171
  definition of, 31
  geometric generation of, 34 ff.
  nowhere differentiability of, 34
  of meander type, 44
  of switch-back type, 44, 46
  similarity dimension of, 159
  three-dimensional, 45
  Wunderlich's versions of, 43 ff.
Peirce, B.O., 133
Peitgen, H.-O., 146, 149, 182
perfect set, 69, 116
period of a *b*-adic representation, 4
Picard, É., 62
Piranian, G., 112, 177, 182
Platzman, L.K., 177
plotting program for the approximating polygons for the Schoenberg curve, 172 ff.
Poincaré, H., 149
Pólya, G., 1, 50, 62, 182
  biography, 62
Pólya's generalization of the Sierpiński-Knopp curve, 62 ff.
polygons, approximating,
  see approximating polygons
positive Jordan content, 1
power series, lacunary, 2, 112
Prat-Villar, Federico, IX, 33
products, infinite, 175

quadratic von Koch curve, 166
quaternaries, 3
  representation of Hilbert curve in terms of, 13 ff.
  representation of Sierpiński-Knopp curve in terms of, 56 ff.
quod erat demonstrandum symbol, 4

Randolph, J.F., 174, 183
range, 4
real numbers, field of, 2
rectifiable curve, 131
reduction ratio, 150
Rham, G. de, 183
Rham curve, 149
Richter, P.H., 149, 182
Robertson, A.G., 109, 183
Rose, N.J., IX, 160, 163
Rosenthal, A., 108, 183
Rudin, W., 121, 183
Russel, B., 31

Sagan, H., 12, 13, 18, 19, 22, 26, 34, 55, 58, 79, 121, 123, 140, 146, 174, 183
Salem, R., 2, 112, 183
Saupe, D., 146, 149, 182
Schaeffer, A.C., 112, 183
Schinzel, A., IX, 86
Schnee, W., 55
Schoenberg, I.J., 2, 62, 119, 123, 183
  biography, 119
Schoenberg's space-filling curve, 119 ff.
  $\aleph_0$ dimensional, 128
  approximating polygons, 123 ff.
  continuity of, 120
  nowhere differentiability of, 121
  surjectivity of, 120
  three-dimensional, 127
Schoenflies, A., 2, 85, 183
Schur, I., 56, 62
Schwarz, H.A., 24
selection theorem of Blaschke, 152
self-similar fractals, 149
set of the excluded middle thirds, 69
sets
  arcwise connected, 117
  compact, 91 ff.
  complements of, 3
  connected, 94 ff.
  dilation of, 150
  disconnected, 94
  dyadic, 116
  invariant attractor, 150, 154 ff.
  locally connected, 98 ff.

sets (*cont.*)
measure of intersection of decreasing sequence of, 174
pathwise connected, 101 ff.
perfect, 69, 116
Sierpiński, W., VII, VIII, 1, 2, 6, 40, 49, 50, 62, 86, 108, 112, 128, 136, 140, 159, 160, 183
biography, 50
Sierpiński
carpet, 159
gasket (triangle), 146, 156
similarity dimension of, 158
sponge, 161
similarity dimension of, 161
triangle (gasket), 146, 156
similarity dimension of, 158
Sierpiński-Knopp curve, 51 ff., 149
approximating polygons for, 60 ff.
as invariant attractor set, 155
as measure preserving map, 111
nowhere differentiability of, 58
Polya's generalization of, 62 ff.
representation in terms of quaternaries, 56 ff.
similarity dimension of, 159
Sierpiński's
modified triangle, 148
Osgood curve, 136
space-filling curve, 50
similarity dimension, 156 ff.
of Cantor set, 158
of Heighway dragon, 164
of Hilbert curve, 159
of von Koch curve, 158
of Mazurkiewicz continuum, 160
of modified Sierpiński triangle, 159
of Peano curve, 159
of Sierpiński curve, 159
of Sierpiński gasket (triangle), 158
of Sierpiński (Menger) sponge, 161
space-filling curve, 5
generation by stochastically independent functions, 108 ff.
Heighway's, 164
Hilbert's, 9 ff.
history of, 1 ff.
Lebesgue's, 69 ff.
Peano's, 31 ff.
Pólya's, 62 ff.
representation by analytic function, 112 ff.
Sierpiński's, 49 ff.
Sierpiński and Knopp's, 51 ff.
Schoenberg's, 119 ff.
sponge, Sierpiński or Menger, 161
square carpet, 159
Stachel, H., 44, 184
Steinhaus, H., 2, 109 ff., 184
Steinitz, E., 149, 184
Steinitz curve, 149
stochastically independent functions, 109 ff.
as coordinate functions of a space-filling curve, 109
as coordinate functions of a measure preserving map, 111
Stromberg, K., 140, 184
Strubecker, K., 11, 13, 184
subcover, finite, 100
surjective map, 3
surjectivity of the
Hilbert curve, 12
Lebesgue curve, 77
Peano curve, 32
Szegö, G., 62

Takagi, T., 166, 184
Takagi curve, 166
ternary, 3
theorems from analysis, 173 ff.
Thomas, E., 140, 181
three-dimensional
Hilbert curve, 26
Peano curve, 45
Schoenberg curve, 127
Titus, C.T., 112, 182
Torhorst, M., 85, 184
topological dimension, 156
trema, 146
triadic representation, 4
Tseng, Shiojenn, 184

uniform distribution functions, 110
uniformly locally connected, 102
unit
cube, 2
interval, 2
square, 2
universal curve, 162
universal plane curve, 161
upper bars denoting periods, 4

Veblen, O., 25, 131, 184
vectors, 3
euclidean norm of, 3
von Koch, see Koch, H. von
Voss, R.F., 146, 165

Walsh, J.L., 133, 184
Weber, H., 24
Weierstraß, K., 5, 6, 24
Weyl, H., 62

Wiener, N., 62
Willard, St., 98, 108, 184
Wirtinger, W., 43
Wirth, N., 170, 172, 184
Wunderlich, W., IX, 13, 18, 21, 44, 149, 185
  biography, 43
Wunderlich's versions of the Peano curve, 44 ff.

Yost, D., 109, 185
Young, G.S., 112, 182

Zeller, K., 56, 180
Zusammenhang im Kleinen, 98
Zygmund, A., 2, 112, 183

Universitext *(continued)*

McCarthy: Introduction to Arithmetical Functions
Meyer: Essential Mathematics for Applied Fields
Meyer-Nieberg: Banach Lattices
Mines/Richman/Ruitenburg: A Course in Constructive Algebra
Moise: Introductory Problem Course in Analysis and Topology
Montesinos: Classical Tessellations and Three Manifolds
Nikulin/Shafarevich: Geometries and Groups
Øksendal: Stochastic Differential Equations
Porter/Woods: Extensions and Absolutes of Hausdorff Spaces
Rees: Notes on Geometry
Reisel: Elementary Theory of Metric Spaces
Rey: Introduction to Robust and Quasi-Robust Statistical Methods
Rickart: Natural Function Algebras
Rotman: Galois Theory
Rybakowski: The Homotopy Index and Partial Differential Equations
Sagan: Space-Filling Curves
Samelson: Notes on Lie Algebras
Schiff: Normal Families of Analytic and Meromorphic Functions
Shapiro: Composition Operators and Classical Function Theory
Smith: Power Series From a Computational Point of View
Smoryński: Logical Number Theory I: An Introduction
Smoryński: Self-Reference and Modal Logic
Stanišić: The Mathematical Theory of Turbulence
Stillwell: Geometry of Surfaces
Stroock: An Introduction to the Theory of Large Deviations
Sunder: An Invitation to von Neumann Algebras
Tondeur: Foliations on Riemannian Manifolds
Verhulst: Nonlinear Differential Equations and Dynamical Systems
Zaanen: Continuity, Integration and Fourier Theory